Title: Unraveling Mysteries: Exploring Science and Debunking Myths - Storytelling for the Whole Family

Author: Ben Chun

Foreword

As a PhD in Chemical Engineering, I've embarked on a voyage through the intricate tapestry of scientific truths and myths, an enlightening and startling journey. My unwavering belief is that science, with all its intricate beauty, must be conveyed in a universally resonating language. This conviction is rooted in a simple observation: when scientific concepts are obscured by complex jargon, they become distant and prone to misunderstanding. Within this gap of understanding, conspiracy theories often find fertile ground.

Out of this need to bridge the gap arises "Unraveling Mysteries: Exploring Science and Debunking Myths - Storytelling for the Whole Family." This book is more than a mere collection of stories; it's a bridge across the chasm of misunderstanding. Here, we simplify the intricate and illuminate the enigmatic. We embark on a journey through the winding paths of enduring science-related conspiracy theories. Each narrative in this collection is a meticulously woven tapestry of truth, not only explaining the origins of these misconceptions but also their systematic and scientific disproof.

My mission extends beyond mere myth-busting. It's about nurturing a profound respect for the scientific method—the rigorous, often painstaking journey to unearth and affirm truths. Understanding this process is paramount to developing a discerning mind—one that questions, probes, and seeks evidence.

As you embark on this journey through the pages of this book, I invite you to join me in a quest for knowledge. Together, we will scrutinize familiar stories through the lens of scientific rigor and rediscover the world around us. This book is a beacon for families, shedding light on the path for young, curious minds contemplating the universe and for adults yearning for a world illuminated by facts, rather than obscured by myths.

Let us journey together, unraveling one mystery at a time, in a world where wonder meets wisdom.

Contents

Foreword

Introduction

Chapter 1: Flat Earth Theory: Dissecting the Cosmic Pancake

Chapter 2: Moon Landing Hoax – Unveiling the Lunar Illusion

Chapter 3: Unraveling the Chemtrail Conspiracy

Chapter 4: Climate Chage Denial

Chapter 5: Confronting Vaccine Misinformation

Chapter 6 Alien Cover-Ups - Unveiling Extraterrestrial Mysteries

Chapter 7: HAARP Weather Control - Unraveling the Conspiracy

Chapter 8: Hollow Earth Theory

Chapter 9: Genetically Modified Organism (GMO) Dangers - Separating Fact from Fiction

Chapter 10: 5G Network Fears – Separating Fact from Fiction

Closing Word

Introduction

Welcome, curious explorers of all ages! Have you ever heard a story so strange and mysterious that it made you wonder if it could be true? In our amazing world, there are countless mysteries and puzzles waiting to be solved. But sometimes, people come up with ideas or stories that sound exciting and even a little bit scary, but they aren't based on real science. These are called conspiracy theories, and they're like puzzles made of make-believe pieces.

I'm a scientist who loves to take apart these make-believe puzzles and show everyone how the real pieces fit together. Think of me as your guide on a thrilling treasure hunt, where instead of looking for buried gold, we're searching for golden nuggets of truth!

In "Unraveling Mysteries: Exploring Science and Debunking Myths - Storytelling for the Whole Family", we'll embark on an adventure across the universe, from the depths of the ocean to the far reaches of outer space. We'll meet these strange stories head-on and use the superpowers of science to see what's true and what's just a tall tale.

Did people really land on the moon? Is the Earth flat like a pancake or round like a basketball? Are there hidden messages in the trails planes leave in the sky? We'll answer these questions and many more!

This book is like a fun-filled science detective story. You don't need to be a science expert to join in; you just need to be curious and ready to explore. So, grab your detective hat, and let's start this exciting journey together. By the end, you'll be a myth-busting superhero, armed with the superpower of knowledge!

Let the adventure begin!

Chapter 1: Flat Earth Theory: Dissecting the Cosmic Pancake

Greetings, fellow explorers of the cosmos! Our journey into the realm of modern conspiracy theories begins with one of the most iconic and perplexing beliefs of all: the Flat Earth Theory. Picture, if you will, the audacious claim that our planet is not the spherical marvel we've come to know through centuries of scientific discovery, but rather a flat expanse, akin to a colossal cosmic pancake.

In this chapter, we shall embark on an expedition into the depths of this captivating theory, unraveling its historical roots and modern resurgence. We shall scrutinize the very essence of this belief, exploring its intriguing premise while keeping our compass firmly grounded in the realm of scientific truth.

Join me as we venture into a world where horizons stretch infinitely, and the laws of physics are put to the test. Through the lens of science, we shall seek to understand why the Flat Earth Theory is scientifically incorrect, uncovering the undeniable proof that our planet is, indeed, a spherical marvel. As our quest unfolds, we will reveal the tools at our disposal to easily disprove this age-old conspiracy, demonstrating the enduring power of scientific inquiry and exploration. So, fasten your seatbelts as we prepare to take off on this cosmic journey, where the quest for truth guides our way!

At the heart of the Flat Earth Theory lies a remarkable assertion: the Earth is not the familiar spherical entity described by astronomers and confirmed through centuries of scientific observation. Instead, proponents of this theory contend that our planet is an expansive, flat plane, stretching endlessly in all directions. To better comprehend this audacious claim, let us delve into the foundational beliefs held by adherents of the Flat Earth Theory:

1. **The Flat Earth Model**: Those who subscribe to this theory often visualize the Earth as a flat, circular disc, surrounded by an immense ice wall that serves as the outermost boundary. In this model, the North Pole lies at the center, while the southern edges are believed to extend indefinitely.

2. **The Sun and Moon**: According to the Flat Earth Theory, the Sun and Moon are not distant celestial objects, but rather luminous discs that revolve above the flat Earth. These luminaries are thought to move in a circular pattern, providing day and night through their motion.

3. **Gravity is a Hoax**: Flat Earthers reject the concept of gravity as described by mainstream science. Instead, they propose that objects fall solely due to the upward motion of the Earth. This idea defies Isaac Newton's universally accepted law of universal gravitation.

4. **Flat Horizon**: Advocates of this theory assert that the horizon always appears flat, regardless of one's altitude or distance from the Earth's surface. They argue that there is no observable curvature, even from high altitudes.

5. **Absence of Antarctica**: In the Flat Earth model, Antarctica is not a continent at the southern pole but rather an encircling ice wall. This notion dismisses the existence of the continent as we know it.

6. **Space is a Deception**: Flat Earthers challenge the concept of outer space, claiming that it is a fabrication created by governments and space agencies to perpetuate the spherical Earth myth.

The Flat Earth Theory, with its imaginative portrayal of our world, has captured the imaginations of many. However, as we delve deeper into this chapter, we will critically examine these claims and assess their validity through the lens of scientific evidence and empirical observations.

Our journey into the intriguing realm of the Flat Earth Theory carries us on a voyage through the corridors of time, uncovering a captivating history steeped in beliefs and misconceptions that have evolved over the centuries. Although the modern resurgence of this theory owes much to the internet and social media, its roots extend far into the annals of human civilization.

In ancient times, several remarkable civilizations embraced the notion of a flat Earth. The Babylonians and Egyptians, among others, envisioned the world as a flat, disc-shaped expanse. Their cosmological views were shaped by limited observations and their interpretations of natural phenomena, paving the way for early iterations of this concept.

As we journey further into history, we encounter the intellectual ferment of ancient Greece. Here, the seeds of spherical Earth understanding were sown by philosophers such as Pythagoras and Parmenides, who proposed that our planet was indeed a sphere. Aristotle, a towering figure in Greek philosophy, added empirical weight to this perspective through his meticulous observations of lunar eclipses and the changing positions of stars as one journeyed from north to south.

However, the narrative of belief in a flat Earth did not wane entirely during the Middle Ages. Some persisted in holding onto this concept, influenced in part by misinterpretations of religious texts. Nevertheless, it is essential to note that educated scholars of the time, influenced by the works of ancient Greeks and Romans, generally recognized the Earth's true spherical nature.

The 19th century brought about a revival of the flat Earth belief with the formation of the Universal Zetetic Society, later known as the Universal Flat Earth Society. At the helm of this movement was Samuel Rowbotham, a key figure who authored the influential work "Zetetic Astronomy: Earth Not a Globe." This marked the inception of the modern Flat Earth movement, rekindling ideas that had largely been discredited by the scientific community.

In our contemporary era, the resurgence of the Flat Earth Theory is intricately tied to the vast expanse of the internet and the connectivity offered by social media. Online communities and YouTube channels dedicated to promoting this theory have gained followers, contributing to its contemporary prominence.

Throughout this historical journey, the Flat Earth Theory has faced the relentless scrutiny of scientific investigation. In the chapters ahead, we will explore the compelling scientific arguments that have not only debunked this theory but also reinforced our understanding of Earth as a magnificent, spherical entity.

What if it were true?

Let us embark on a journey into the heart of imagination, where the Flat Earth Theory reigns supreme. If, by some twist of fate, this theory were indeed true, the consequences would be nothing short of revolutionary:

In this alternate reality, our understanding of the cosmos would undergo a radical transformation. The very foundations of astronomy, physics, and geography would be shaken to their core, giving rise to an entirely different narrative of the universe.

Picture a world where the Earth is not the familiar spherical marvel but instead a flat, circular disc. In this scenario, the North Pole stands at the center, while the southern edges stretch out indefinitely. This fundamental reshaping of our planetary perception would challenge the way we perceive our place in the universe.

The Sun and Moon, which we currently regard as distant celestial objects, would no longer occupy their traditional roles. Instead, they would be reimagined as luminous discs that gracefully traverse the sky above the flat Earth. The daily cycle of day and night would unfold as these celestial discs made their circular journey, a far cry from the astronomical reality we accept today.

In the world of the Flat Earth Theory, gravity as we know it would be but a fanciful notion. The laws governing motion and mass, established by Isaac Newton and later refined by Albert Einstein, would be rendered obsolete. Objects would not fall towards the Earth due to gravitational attraction; instead, they would rise upwards at a constant rate, defying the very essence of our understanding of physics.

Imagine a horizon that always appears flat, regardless of one's altitude or distance from the Earth's surface. Mountains, skyscrapers, and distant landscapes would defy the laws of perspective as we currently understand them. This notion starkly contrasts with the observed curvature of the Earth from high altitudes and across vast expanses of land and sea.

In this theoretical construct, the traditional continent of Antarctica would vanish as we know it, replaced by an encircling ice wall. This ice wall would serve as the outermost boundary of the flat Earth, challenging our perceptions of geography and exploration in the polar regions.

Furthermore, proponents of the Flat Earth Theory assert that outer space is not the vast cosmos we imagine but an elaborate fabrication, conjured by governments and space agencies to maintain the illusion of a spherical Earth. Astronaut missions, space exploration, and our comprehension of the broader universe would all be cast into doubt.

However, it is imperative to recognize that this scenario of a flat Earth stands in stark contrast to an overwhelming body of scientific evidence and centuries of empirical observations. In the upcoming chapters, we shall delve into the robust scientific refutations that disprove this theory and solidify our understanding of Earth as a magnificent, spherical entity.

Why It's Scientifically Incorrect

Our quest to understand the captivating enigma of the Flat Earth Theory brings us face to face with a profound contradiction. This theory, although alluring in its audacity, is fundamentally incongruent with the bedrock of scientific knowledge and centuries of empirical evidence. Let us venture into the realm of scientific exploration to fathom why the Flat Earth Theory falters when scrutinized through the lens of established scientific principles.

Gravity, a fundamental force that governs the motion of objects in our universe, forms the cornerstone of our understanding of the natural world. It was elucidated by the genius of Isaac Newton and later expanded upon by Albert Einstein. This force explains why objects fall toward the Earth, how celestial bodies remain in graceful orbits, and why we remain anchored to the planet's surface. The Flat Earth Theory's dismissal of gravity, instead proposing that objects are propelled upwards, challenges the very essence of these well-substantiated principles, devoid of any empirical evidence to support such a remarkable claim.

One of the most conspicuous pieces of evidence for Earth's spherical nature is the observable manner in which distant objects, such as ships at sea, gradually vanish from view. As a vessel sails away from an observer, it appears to sink below the horizon due to the Earth's curvature. This phenomenon, consistently witnessed across the globe, stands as a tangible testament to our planet's spherical form.

Further compelling proof emerges during lunar eclipses when the Earth casts its shadow on the Moon's surface. The shadow, consistently observed to be round, directly contradicts the flat Earth model. This circular shadow is a direct consequence of the Earth being positioned between the Sun and the Moon, a configuration that is only feasible if the Earth is indeed a sphere.

Elevated locations, including mountains, skyscrapers, and airborne vantage points, grant us the privilege of gazing upon distant landscapes and horizons that consistently appear curved. This curvature is a direct consequence of the Earth's spherical shape and is repeatedly confirmed through countless observations around the world.

The practical ability to circumnavigate the globe by traveling in a straight line and returning to the starting point serves as an unequivocal demonstration of the Earth's spherical nature. This feat, accomplished by countless explorers and adventurers, would be an impossibility if the Earth were flat without turning back at some point.

Over the past few decades, astronauts aboard spacecraft and satellites have captured a myriad of awe-inspiring images of the Earth from space. These photographs, portraying our planet as a breathtaking blue sphere, provide irrefutable visual confirmation of Earth's spherical form.

Finally, the precision and efficiency of aircraft flight paths across the globe are grounded in the concept of a spherical Earth. Flight routes, including the ability to traverse shorter paths over the polar regions, rest upon the curvature of our planet's surface as validated by empirical evidence.

In essence, the Flat Earth Theory finds itself in direct contradiction with an extensive body of scientific knowledge forged and corroborated over centuries. It dismisses the principles of gravity, disregards observable phenomena such as the curvature of the horizon, and fails to offer credible alternative explanations. As we embark on the subsequent chapters that delve deeper into the scientific refutations of this theory, a resounding truth will emerge: the Earth stands resolutely as a spherical marvel, an incontrovertible reality substantiated by an overwhelming body of scientific evidence and empirical observations.

As we embark on our quest to dismantle the enigma of the Flat Earth Theory, we set forth to illuminate the glaring contradictions that render this theory untenable in the face of scientific scrutiny. Our journey into debunking this theory takes us through a landscape of concrete experiments, irrefutable observations, and steadfast principles, all of which converge to underscore the indisputable fact of our planet's spherical nature.

First and foremost, we encounter the ship's vanishing act, a phenomenon witnessed by countless mariners and coastal observers. As ships sail away from shore, they gradually vanish from view, with their lower portions disappearing first. This gradual disappearance is a direct consequence of the Earth's curvature, a fundamental aspect of our world that has been consistently observed and documented throughout history.

Next, we navigate the intricate web of global flight paths, where the efficiency and precision of air travel routes depend on the curvature of the Earth. The ability to fly non-stop between distant continents, such as the direct flight from New York to Sydney, hinges on the globe's spherical shape. These flight routes circumvent the need for meandering detours that would be essential in a flat Earth scenario.

Ascending to higher altitudes, we find ourselves face to face with the horizon's gentle curve. Whether standing on the shoreline, ascending a towering mountain, or boarding an airplane, our vantage point allows us to witness the gradual curvature of the horizon. This observable phenomenon underscores the spherical nature of our world, as it extends our line of sight farther than a flat plane ever could.

In the celestial realm, lunar eclipses offer a compelling spectacle that directly contradicts the Flat Earth Theory. During these rare events, the Earth casts a distinct, round shadow upon the Moon's surface. This circular shadow aligns precisely with the anticipated shadow produced by a spherical Earth positioned between the Sun and the Moon.

The concept of gravity, as articulated by the laws of physics, relies upon the Earth's spherical form. The gravitational force exerted by a spherical mass at its center underpins the phenomenon of weight, which varies depending on one's location on Earth. Flat Earth models falter in providing a coherent explanation for the consistent observations of varying weights across the globe.

Finally, the wealth of images and videos captured by astronauts and satellites orbiting our planet serves as an unassailable testament to the Earth's true form. These visual records depict the Earth as a stunning, spherical entity suspended in the cosmos, reinforcing the irrefutable reality of our world.

In the face of these compelling experiments, observations, and principles, the Flat Earth Theory crumbles like a sandcastle before the tide. It is powerless to provide a coherent, evidence-based alternative to the centuries of scientific understanding that have firmly established our planet's spherical shape. The unwavering evidence derived from these empirical observations leaves no room for doubt: the Earth, our celestial abode, is undeniably a magnificent sphere, firmly ensconced in the realm of scientific fact.

Conclusion

Our exhilarating expedition into the captivating realm of the Flat Earth Theory draws to a close, leaving us with a profound revelation—one that celebrates the enduring power of scientific inquiry and the unwavering quest for truth. In our journey, we have traversed the landscapes of belief and skepticism, exploring a theory that challenges the very foundations of our understanding of the world.

The Flat Earth Theory, with its imaginative portrayal of a flat, disc-shaped Earth, has captured the imaginations of many throughout history and into the present day. From ancient civilizations to the modern resurgence of this belief on the internet, the theory has persisted as a testament to human curiosity and the enduring allure of the unknown.

However, as we have ventured deeper into the heart of this theory, we have encountered a resounding truth—a truth anchored in centuries of scientific discovery, empirical evidence, and the cumulative wisdom of humanity. The evidence supporting the spherical nature of our planet is overwhelming, from the ship's gradual disappearance on the horizon to the curvature of the Earth observed from elevated vantage points and the compelling imagery captured from space.

Our exploration has reaffirmed the fundamental principles of science and the importance of critical thinking. It has underscored the invaluable role of empirical observation and the scientific method in our pursuit of knowledge. The Flat Earth Theory, while captivating in its audacity, ultimately crumbles in the face of this unwavering scientific consensus.

As we return from our odyssey through belief and skepticism, we carry with us the torch of reason and enlightenment. We have unraveled the mysteries and debunked the myths, emerging as myth-busting superheroes armed with the superpower of knowledge. Our journey reminds us that in a world brimming with mysteries and puzzles, the pursuit of truth through the lens of science remains our most potent guide.

So, let us embark on further expeditions, exploring the wonders of our universe, unraveling the intricacies of nature, and unveiling the secrets of existence. For in the end, it is our relentless pursuit of knowledge that illuminates the path to a brighter, more informed future. The adventure continues, ever onward.

Chapter 2: Moon Landing Hoax – Unveiling the Lunar Illusion

Welcome to the chapter dedicated to unraveling the enigmatic Moon Landing Hoax conspiracy theory. In the vast tapestry of skepticism and intrigue that surrounds the moon landings, we find ourselves at a crossroads of history and curiosity. The claim that the Apollo moon landings were an elaborate Hollywood-style fabrication has endured for decades, casting a shadow over one of humanity's most iconic achievements.

In this journey, we shall embark on a mission to sift through the claims and counterclaims, the evidence and skepticism, and ultimately, the scientific and empirical truth. It is here, in the chapters to come, that we shall examine the moon landings with the scrutiny they deserve, using the superpowers of science and critical thinking to separate the lunar illusion from reality.

So, fasten your seatbelts and don your astronaut helmets, for our adventure begins now, as we explore the captivating realm of the Moon Landing Hoax.

In our quest to unveil the truth behind the Moon Landing Hoax, we must first confront the central assertion of this intricate conspiracy theory: the belief that the United States government, in a bid to assert its dominance during the Cold War era, staged the moon landings as an elaborate ruse. At the heart of this claim lies the audacious notion that the iconic Apollo missions, which saw astronauts setting foot on the lunar surface, were not genuine feats of human exploration, but rather Hollywood-style productions filmed right here on Earth.

Proponents of this theory argue that the motives behind such a hoax were manifold. They contend that the United States sought to bolster its global standing during the Cold War by asserting superiority over its rival, the Soviet Union. In their view, faking a moon landing would serve as a propaganda coup and a demonstration of technological prowess, effectively bolstering American prestige on the world stage.

This belief in a lunar hoax hinges on the perception of the moon landings as a grand theatrical performance. Advocates of the conspiracy theory cite various elements they deem suspicious, from anomalies in photographs to questions about the authenticity of the moon rock samples. They argue that these perceived inconsistencies point to a meticulous effort to maintain the illusion.

As we delve deeper into this chapter, we shall meticulously examine the historical context in which the moon landings occurred, the significance of the Apollo program, and the geopolitical backdrop against which it unfolded. By understanding the era in which these remarkable achievements took place, we can better assess the validity of the Moon Landing Hoax conspiracy and determine whether the evidence supports or debunks this audacious claim.

To fully grasp the backdrop against which the Moon Landing Hoax conspiracy theory emerged, we must venture into the historical context of the 1960s—a tumultuous era marked by intense geopolitical rivalry and technological innovation.

The Cold War, a protracted struggle for global dominance between the United States and the Soviet Union, cast a long shadow over this period. The competition between these two superpowers extended beyond military might and political influence; it reached into the uncharted realms of space. The launch of the Soviet satellite Sputnik in 1957 had already ignited the space race, with both nations vying for supremacy in the cosmos.

It was within this crucible of competition and Cold War tension that the Apollo program, conceived by President John F. Kennedy in 1961, took shape. Kennedy's vision of landing an American astronaut on the moon before the end of the decade set in motion a series of missions that would come to define an era. The stakes were high, with both the United States and the Soviet Union racing to demonstrate their technological prowess and ideological superiority.

The Apollo program, a colossal undertaking that combined the efforts of thousands of scientists, engineers, and astronauts, represented the pinnacle of American scientific and engineering achievement. Its missions aimed not only to explore the lunar surface but also to push the boundaries of human knowledge and capability. These endeavors captured the imagination of the world and became emblematic of human potential.

Against the backdrop of the Cold War and the fierce competition between superpowers, the moon landings held profound symbolic significance. They were seen as a testament to American ingenuity and resilience, a reflection of the nation's unwavering commitment to achieving the seemingly impossible. It was against this historical canvas that the conspiracy theory of a moon landing hoax took root, as skeptics questioned the authenticity of these monumental achievements.

In the chapters to come, we shall delve deeper into the historical context surrounding the moon landings and assess the veracity of the Moon Landing Hoax conspiracy in light of the geopolitical pressures and technological advancements of the time. Our journey will continue as we separate fact from fiction and explore the enduring legacy of the Apollo program.

The Apollo Missions

Our voyage into the heart of the Moon Landing Hoax conspiracy invites us to explore the rich tapestry of the Apollo missions, each an epic chapter in the grand narrative of human lunar exploration.

Apollo 11: Our odyssey commences with Apollo 11, launched on July 16, 1969, with astronauts Neil Armstrong, Buzz Aldrin, and Michael Collins on board. The lunar module "Eagle" descended to the moon's surface on July 20, 1969, touching down in the Sea of Tranquility. Neil Armstrong's first historic steps were not the only memorable moments of this mission. Buzz Aldrin's unique lunar footprint left an impression not just in the soil but also in the annals of history. Fun fact: Armstrong's descent to the surface was perilous, with fuel running low and a field of boulders in their path, but his exceptional piloting skills ensured a safe landing with just seconds to spare.

Apollo 12: Mere months later, in November 1969, Apollo 12 embarked on its lunar landing mission. Commanded by Charles "Pete" Conrad and piloted by Alan L. Bean, the lunar module "Intrepid" touched down within walking distance of the Surveyor 3 spacecraft, a previous robotic mission. This precision landing was no accident; it was a testament to NASA's growing mastery of lunar exploration. Fun fact: While on the lunar surface, Alan Bean inadvertently pointed the television camera at the sun, temporarily blinding it. Mission control instructed them to turn it off and on again, famously solving the problem.

Apollo 13: The Apollo 13 mission, launched in April 1970, was destined to become one of the most harrowing chapters in space exploration history. Astronauts James Lovell, John Swigert, and Fred Haise faced a life-threatening situation when an oxygen tank exploded en route to the moon. The mission had to be aborted, and the astronauts conducted a "slingshot" trajectory around the moon to return safely to Earth. Fun fact: Despite the life-threatening situation, humor prevailed in space, with Lovell famously communicating, "Houston, we've had a problem here." The phrase was later popularly misquoted as "Houston, we have a problem."

Apollo 14-17: Subsequent Apollo missions, including Apollo 14-17, continued to expand the boundaries of lunar exploration. Apollo 14, led by Alan Shepard and Edgar D. Mitchell, introduced a delightful twist to lunar activities. Shepard, an avid golfer, famously brought a makeshift six-iron golf club and hit golf balls on the lunar surface. It was a literal "moonshot" that traveled surprisingly far due to the moon's lower gravity. Fun fact: Shepard noted that his swing was less restricted by his spacesuit than anticipated, and the golf balls went further than expected.

Scientific Achievements: The Apollo missions were not just moonwalks; they were substantial scientific endeavors. Astronauts conducted experiments, deployed scientific instruments, and collected rock and soil samples. These lunar rock samples, numbering in the hundreds of kilograms, have since been studied extensively, providing invaluable insights into the moon's geological history and its connection to Earth.

Global Acknowledgment: The moon landings were not just an American triumph; they were a global celebration of human ingenuity and exploration. Radio broadcasts, television coverage, and print media conveyed the historic nature of these missions to millions, leaving an indelible mark on the collective human consciousness and symbolizing the boundless potential of human endeavor.

As we journey further into this chapter, we shall continue to explore the scientific and empirical evidence that firmly establishes the authenticity of the moon landings, solidifying the foundation for a comprehensive debunking of the Moon Landing Hoax conspiracy theory.

Our journey through the Moon Landing Hoax conspiracy now takes us to the heart of the matter: the comprehensive debunking of the captivating claims that have challenged the authenticity of the Apollo moon landings. These monumental achievements in human history have faced relentless scrutiny and skepticism by conspiracy theorists, prompting us to dissect these allegations and unveil the scientific and empirical evidence that unequivocally upholds the reality of the moon landings.

One of the frequently cited pieces of evidence by moon landing skeptics revolves around alleged anomalies in the photographs captured during the missions. Claims of inconsistent shadows, unusual lighting, reflections, and background discrepancies are often presented as proof of a hoax. However, these contentions often stem from a misunderstanding of lunar photography techniques and the unique lighting conditions found on the moon's surface.

The myth of the "waving" American flag planted on the lunar surface is another iconic claim put forth by skeptics. According to their arguments, the flag's appearance of movement in the vacuum of space is deemed impossible. In reality, this visual effect is attributed to the absence of air resistance and the momentum imparted to the flag during its insertion into the lunar soil.

Radiation hazards posed by the Van Allen radiation belts surrounding the Earth have also raised concerns among moon landing skeptics. Critics assert that astronauts could not have survived passage through these belts. However, NASA meticulously planned the trajectories of the Apollo missions to minimize radiation exposure, and the relatively short duration of passage through the belts posed no significant threat to the astronauts.

One of the most compelling pieces of evidence in support of the moon landings comes in the form of retroreflectors left on the lunar surface by the Apollo 11, 14, and 15 missions. These retroreflectors continue to bounce back laser beams sent from Earth, serving as a tangible testament to human presence on the moon.

The authenticity of the moon landings has also been independently corroborated by multiple sources. The Soviet Union, during the height of the Cold War and America's space race rival, closely monitored the Apollo missions and would have been quick to expose any hoax. Furthermore, amateur radio operators from around the world tracked the missions and verified the authenticity of the broadcasts.

Scores of scientists, engineers, and experts from various fields have provided their testimonies in support of the moon landings. Their collective expertise forms a formidable barrier to moon landing deniers. Experts in disciplines such as photography, geology, physics, and aerospace engineering have independently reviewed and validated the data and evidence.

As we venture deeper into this chapter, we shall systematically address these claims and provide scientific explanations that unravel the inaccuracies and misconceptions behind the Moon Landing Hoax conspiracy theory. The evidence, accumulated over decades of meticulous research and exploration, reaffirms the authenticity of the moon landings and underscores the unwavering truth of humanity's extraordinary journey to the lunar surface.

The Role of Technology and Science

In our unwavering pursuit of truth regarding the Moon Landing Hoax, we now turn our gaze towards the pivotal role played by technology and science in the Apollo missions. These extraordinary achievements in lunar exploration were not merely grand theatrical productions but intricate, meticulously planned endeavors grounded in cutting-edge science and technology.

Rocketry and Propulsion: At the core of the Apollo program's success lay the mastery of rocketry and propulsion. The colossal Saturn V rocket, standing at a towering 363 feet, remains one of the most powerful machines ever built. Its brute force propelled astronauts towards the moon at astounding speeds, far beyond the capabilities of any terrestrial aircraft. The physics and engineering underpinning these mammoth rockets are unassailable, serving as a testament to human ingenuity.

Navigation and Guidance: Precise navigation and guidance were indispensable for lunar missions. Advanced computers and inertial navigation systems calculated trajectories with pinpoint accuracy. Astronauts used star sightings, celestial landmarks, and onboard guidance systems to ensure their spacecraft stayed on course. These technologies formed the bedrock of safe lunar landings.

Life Support and Suits: The lunar environment is an unforgiving vacuum, devoid of atmosphere and subject to extreme temperature fluctuations. Space suits, constructed with multiple layers of specialized materials, provided life support and protection. The backpack-like Portable Life Support System (PLSS) enabled astronauts to breathe, maintain temperature, and communicate while exploring the lunar surface. These suits were a triumph of engineering, debunking claims of their impracticality.

Lunar Modules: The lunar modules themselves, used for descent and ascent from the moon's surface, were marvels of engineering. The Lunar Module (LM) featured a unique design tailored to the lunar environment, with legs designed for soft landings and an ascent stage for return to the command module. The functionality and purpose of these spacecraft have been corroborated by meticulous engineering analyses.

Lunar Geology and Samples: Apollo astronauts conducted experiments and collected rock and soil samples from the moon's surface. These lunar samples, numbering in the hundreds of kilograms, have been subject to rigorous scientific study by experts worldwide. The insights derived from these samples have provided invaluable information about lunar geology and its relation to Earth's history.

The technologies and scientific principles underpinning the Apollo missions represent a culmination of human achievement. To contend that such a colossal undertaking could have been orchestrated as a mere hoax disregards the wealth of empirical evidence, the testimonies of experts, and the enduring legacy of space exploration. As we progress further into this chapter, we shall continue to uphold the integrity of these achievements and counter claims that undermine the significance of these scientific advancements.

In our quest to debunk the Moon Landing Hoax, we arrive at a pivotal juncture: the enduring legacy of the Apollo program. Beyond the confines of historical records and empirical evidence, the Apollo missions have left an indelible mark on human history and continue to shape our exploration of space.

The Apollo missions were not limited to planting flags and collecting rocks; they expanded the boundaries of human knowledge. The lunar samples returned to Earth have yielded invaluable insights into the moon's geological history, the composition of its surface, and even clues about the early solar system. These discoveries continue to inform our understanding of planetary science.

Moreover, the legacy of Apollo lives on in modern lunar exploration efforts. Robotic missions, such as the Lunar Reconnaissance Orbiter (LRO) and China's Chang'e missions, have mapped the moon's surface, identified potential landing sites for future missions, and provided a new perspective on our celestial neighbor. These missions build upon the foundation laid by Apollo.

Furthermore, the Apollo program ignited the imaginations of countless individuals, inspiring generations of scientists, engineers, and explorers. It has been a driving force behind STEM (Science, Technology, Engineering, and Mathematics) education and continues to motivate young minds to pursue careers in space exploration and related fields.

The spirit of international collaboration that began with Apollo endures in endeavors like the International Space Station (ISS), where multiple nations come together to conduct scientific research and demonstrate peaceful cooperation in space. This cooperative spirit is a testament to what humanity can achieve when nations work together.

The vision of human exploration beyond low Earth orbit, encapsulated in the Artemis program and aspirations for lunar colonization, is an extension of the Apollo legacy. These ambitions are driven by a deep-seated belief in the potential for human exploration to expand our scientific horizons and pave the way for future missions to Mars and beyond.

The Apollo program, far from being a historical relic, remains a vibrant force in shaping the future of space exploration. Its scientific discoveries, inspirational impact, and collaborative spirit continue to influence our journey into the cosmos. As we delve further into this chapter, we will underscore the ongoing significance of the Apollo legacy and reaffirm the unwavering truth of humanity's remarkable journey to the lunar surface.

Conclusion

As we reach the culmination of our exploration into the Moon Landing Hoax conspiracy, it becomes abundantly clear that the moon landings were not a clandestine production staged on Earth but a monumental achievement of human exploration. Our journey through this chapter has illuminated the steadfast evidence and irrefutable facts that debunk the claims of skeptics.

The scientific community, comprised of experts from various fields, has provided unequivocal support for the authenticity of the moon landings. Their meticulous analyses of lunar samples, photography, engineering, and data have left no room for doubt. The enduring legacy of the Apollo program, which continues to shape space exploration today, stands as a testament to its authenticity.

We've explored the technology, science, and expertise that underpinned the missions, dispelling misconceptions and demonstrating the sheer impossibility of orchestrating a hoax of this magnitude. Moreover, the enduring legacy of Apollo, from inspiring future generations of scientists to fostering international collaboration, cements its place in history.

In our journey, we've learned that skepticism, while a crucial part of the scientific process, must be grounded in facts and evidence. The claims of the Moon Landing Hoax conspiracy, when subjected to scrutiny, crumble in the face of overwhelming proof. The moon landings are not just a part of American history; they are a testament to human achievement and the relentless pursuit of knowledge.

As we conclude this chapter, let us celebrate the incredible achievements of the Apollo program and continue to look to the stars with wonder and curiosity. The moon landings are not a tale of deception but a testament to the power of human ingenuity and the enduring quest to explore the cosmos. Embrace the lunar truth, for it is a shining beacon of what humanity can accomplish when we reach for the stars.

Chapter 3: Unraveling the Chemtrail Conspiracy

Welcome to the heart of the chemtrail conspiracy theory—a realm where white trails left by aircraft are believed to be more than just innocent condensation trails. In this subchapter, we dive deep into the core claims that fuel the chemtrail conspiracy, understanding the belief that these trails are, in fact, chemical sprays and exploring the purported motivations behind what some allege to be a covert operation.

The chemtrail conspiracy revolves around the notion that the white streaks left behind by aircraft, often called contrails (short for condensation trails), are something far more sinister. Instead of being formed by the condensation of water vapor in the aircraft's exhaust, proponents of the conspiracy claim that these trails consist of chemicals deliberately released into the atmosphere. This theory posits that governments or shadowy organizations are engaged in a covert operation to disperse chemicals for various, often nefarious, purposes.

Proponents of the chemtrail theory allege that the composition of these supposed chemical trails varies, with claims ranging from harmful pollutants to mind-altering substances. Commonly cited elements include aluminum, barium, strontium, and even biological agents. The conspiracy asserts that these chemicals are being dispersed at high altitudes, posing health risks to the population below.

The motivations attributed to those behind the alleged chemtrail program are diverse and often speculative. While claims vary, some recurring themes include weather modification, population control, mind control, environmental manipulation, or military experimentation. The conspiracy paints a picture of shadowy figures with ulterior motives, working diligently to manipulate our skies and, by extension, our lives.

To comprehensively address the chemtrail conspiracy, it is essential to delve into its historical roots and trace the evolution of this perplexing theory. The development of the chemtrail conspiracy has been influenced by historical events, government actions, and societal shifts, all of which have contributed to its growth and persistence.

The origins of the chemtrail conspiracy theory can be traced back to the mid-20th century. It emerged as an offshoot of earlier conspiracy theories related to aerial spraying, which often alleged the release of chemical agents for various purposes, such as weather modification or population control.

A significant catalyst for the growth of the chemtrail conspiracy was the historical context of government secrecy and mistrust. The Cold War era, marked by covert military operations and classified projects, nurtured a climate of suspicion. This environment led some individuals to question official narratives and cultivate a deep-seated skepticism of government actions.

Throughout its development, the chemtrail conspiracy theory has been shaped by key events and influential figures. Notable incidents, such as military experiments involving chemical agents, have provided fuel for conspiracy narratives. Likewise, charismatic individuals and groups have played pivotal roles in popularizing and disseminating the conspiracy theory.

What If It's True?

Within the realm of conspiracy theories, it is imperative to explore the hypothetical scenarios posed by the chemtrail conspiracy if it were to be considered true. By investigating the potential consequences of the chemtrail conspiracy theory, we can better assess its validity and the gravity of its implications.

If the chemtrail conspiracy were indeed accurate, the implications would be far-reaching. Proponents claim that the release of chemical agents into the atmosphere has profound effects on both human health and the environment. Investigating these claimed consequences is crucial to understanding the theory's impact.

One of the central concerns in the chemtrail conspiracy is the potential harm posed to human health. Proponents suggest that exposure to the chemicals dispersed via chemtrails can lead to a range of health issues, including respiratory problems, neurological disorders, and even long-term chronic illnesses. Investigating these claims entails assessing the scientific validity of the alleged health risks.

Beyond health concerns, the chemtrail conspiracy posits broader environmental and geopolitical implications. These may include disruptions to ecosystems, alterations in weather patterns, or even manipulation of global events. Evaluating the feasibility of such assertions and their potential impact on our planet is essential to gauging the validity of the theory.

Debunking the Chemtrail Conspiracy

In our relentless pursuit of truth and scientific understanding, we now confront the central claims of the chemtrail conspiracy theory with an unyielding focus on empirical evidence and established scientific principles. This subchapter is dedicated to systematically dismantling the assertions made by proponents of the theory and shedding light on the facts that debunk the chemtrail conspiracy.

To dispel the chemtrail conspiracy, it is paramount to comprehend the science behind contrails—those white streaks in the sky that form the basis of the theory. Contrails, short for condensation trails, are a natural byproduct of aircraft engine exhaust. They are composed primarily of water vapor, which freezes at high altitudes due to the frigid temperatures.

Central to the chemtrail conspiracy are numerous misconceptions about contrails. Critics often point to the persistence of contrails as evidence of chemical spraying, but this phenomenon can be explained by atmospheric conditions. The role of temperature, humidity, and altitude in contrail formation will be elucidated to clarify these misconceptions.

One of the most compelling arguments against the chemtrail conspiracy is the wealth of empirical evidence and expert opinions. Researchers, atmospheric scientists, and aviation experts have rigorously studied contrails and have consistently affirmed their natural origin. We will explore their findings and insights, which provide an unshakable foundation of evidence.

To gain a more profound understanding of the chemtrail conspiracy's scientific foundation, we must embark on an in-depth exploration of contrails—those enigmatic white trails left by aircraft in the sky. By dissecting the formation process, composition, and characteristics of contrails, we can demystify these natural phenomena and distinguish them from the alleged chemical trails posited by the conspiracy theory.

Contrails, short for condensation trails, are formed when the exhaust gases from aircraft engines come into contact with extremely cold temperatures at high altitudes. These gases contain water vapor, which, upon exposure to the frigid upper atmosphere, undergoes rapid condensation and freezing. This process results in the creation of the visible white streaks that appear behind aircraft.

The persistence and characteristics of contrails are intrinsically linked to atmospheric conditions. Key factors include temperature, humidity, and altitude. Understanding how these variables influence contrail formation and dissipation is crucial to debunking the notion that all visible trails in the sky are chemtrails.

Contrails consist primarily of tiny ice crystals, which collectively contribute to their white appearance. These ice crystals are composed of frozen water vapor, and their composition aligns with the expected byproducts of aircraft engine exhaust. By examining the fundamental composition of contrails, we can affirm their natural origin.

Drawing a clear distinction between contrails and chemtrails is essential to refuting the conspiracy theory. Contrails, as scientifically understood, are composed of naturally occurring substances consistent with aircraft engine emissions. This differentiation will help dispel misconceptions and misidentifications in the sky.

The Role of Aviation and Atmospheric Science

As we continue our scientific journey to debunk the chemtrail conspiracy, we turn our attention to the invaluable contributions of aviation and atmospheric science. These fields provide critical insights into the dynamics of contrail formation, the composition of aircraft emissions, and the feasibility of a large-scale chemtrail program. By examining the rigorous research conducted by experts in these domains, we further bolster the case against the conspiracy.

Aviation experts have played a pivotal role in unraveling the mysteries of contrails. Their research encompasses a comprehensive understanding of aircraft engines, emissions, and contrail formation. By scrutinizing their findings, we gain a deep appreciation of the scientific underpinnings of contrails and their distinction from chemtrails.

One of the compelling aspects of debunking the chemtrail conspiracy is the transparency of flight data. Flight records, emissions data, and aircraft specifications are accessible to the public and undergo rigorous scrutiny by aviation authorities. This transparency ensures accountability and reinforces the authenticity of contrails as natural phenomena.

Critics of the chemtrail conspiracy contend that the scale and logistical complexity of such a program would be insurmountable. Examining the practicality of dispersing chemicals from commercial aircraft at high altitudes offers further insights into the implausibility of the theory. Aviation and atmospheric science experts provide valuable perspectives on the feasibility of covert chemical spraying.

In our journey to uncover the truth behind the chemtrail conspiracy, it is imperative to recognize the broader societal implications of such unfounded claims. Misinformation, especially in the age of the internet and social media, can have far-reaching consequences that extend beyond the realm of science. This subchapter explores the societal impact of the chemtrail conspiracy and underscores the importance of responsible discourse and critical thinking.

The chemtrail conspiracy, like many other unfounded theories, has the potential to sow seeds of doubt and distrust in society. As the theory gains traction, it can erode public confidence in institutions, government agencies, and scientific expertise. Such erosion of trust can have profound consequences for informed decision-making and public policy.

The prevalence of conspiracy theories underscores the vital importance of media literacy and critical thinking skills. In an era of rapid information dissemination, individuals must be equipped to discern credible sources from misinformation. Encouraging these skills is essential for safeguarding against the undue influence of baseless conspiracies.

Promoting responsible discourse is an ethical imperative. Fact-checking, scrutinizing claims, and seeking reliable sources are essential steps in combating the spread of misinformation. It is incumbent upon individuals, educators, and media outlets to prioritize the dissemination of accurate information and the rejection of baseless conspiracies.

Ultimately, our journey through the chemtrail conspiracy reinforces the value of the pursuit of truth. In the face of unfounded claims, the scientific method, empirical evidence, and rigorous research remain unwavering tools for discerning fact from fiction. Embracing the pursuit of truth is an affirmation of our commitment to knowledge and the rejection of misinformation.

Conclusion

As we reach the culmination of our scientific expedition into the realm of the chemtrail conspiracy, the verdict becomes unequivocal: the claims of this theory, positing that the white trails left by aircraft are chemical sprays with sinister intentions, crumble under the weight of scientific evidence and critical scrutiny. In this concluding subchapter, we recapitulate the key findings and scientific facts that dispel the chemtrail myth and affirm the authenticity of contrails as natural atmospheric occurrences.

Our journey through this chapter has underscored the paramount role of scientific understanding in debunking conspiracy theories. By delving into the science of contrails, we have dismantled the central claims of the chemtrail conspiracy, highlighting the natural processes that give rise to these phenomena.

Addressing misconceptions and mistrust is a pivotal aspect of confronting conspiracy theories. The chemtrail conspiracy, born from suspicion and misinformation, serves as a stark reminder of the importance of transparent communication, scientific education, and critical thinking in combating unfounded beliefs.

Throughout our exploration, we have witnessed the power of empirical evidence and expert consensus. The collective body of research by aviation and atmospheric science experts, alongside the transparency of flight data, leaves no room for doubt regarding the authenticity of contrails as naturally occurring phenomena.

The chemtrail conspiracy also highlights the imperative of responsible discourse and fact-checking in an age characterized by rapid information dissemination. Society's resilience against misinformation hinges on its collective commitment to upholding the standards of credible sources and rigorous verification.

In our journey through the chemtrail conspiracy, we have unveiled the scientific truths that underpin the authenticity of contrails and debunk the baseless claims of the theory. Let this chapter serve as a testament to the power of knowledge, critical thinking, and scientific inquiry in dispelling myths and embracing the pursuit of truth. As we move forward, let us champion the values of scientific literacy, media literacy, and responsible discourse, ensuring that our collective pursuit of knowledge remains unwavering in the face of unfounded beliefs.

Chapter 4: Climate Chage Denial

Welcome to the heart of the climate change denial debate, where we confront a perplexing phenomenon: despite an overwhelming scientific consensus, some individuals assert that climate change is either a complete hoax or grossly exaggerated. In this subchapter, we embark on a deep dive into the core claims that fuel climate change denial, seeking to understand the foundations of this skepticism and the motivations behind it.

Climate change denial encompasses a spectrum of beliefs that cast doubt on the reality, causes, and severity of global climate change. At one end of the spectrum, some deny the existence of climate change altogether, asserting that variations in temperature and weather patterns are natural and not influenced by human activities. At the other end, some accept the existence of climate change but argue that it is exaggerated for political or economic gain.

Among the central claims of climate change denial is the assertion that climate change is a complete hoax. Deniers argue that scientists and institutions worldwide have conspired to manufacture a crisis for personal or political motives. They often cite instances of errors or controversies in climate science as evidence of a deliberate deception.

Another facet of climate change denial involves allegations that the severity of climate change is grossly exaggerated. Deniers argue that the severity of climate change is grossly exaggerated. Deniers argue that scientists and policymakers have inflated the potential impacts of climate change to secure research funding, enact regulations, or advance particular agendas. They may downplay the significance of climate models and scientific consensus as tools for fearmongering.

Understanding the motivations behind climate change denial is crucial in unraveling this complex phenomenon. Motives vary among individuals and groups, encompassing economic interests, political ideologies, and a general resistance to change. Some may view climate action as a threat to certain industries, while others perceive it as an encroachment on personal freedoms or a challenge to their worldview.

To comprehend the persistence of climate change denial, we must embark on a historical journey that unveils the roots and evolution of this skepticism. Understanding the historical context is crucial in illuminating the societal, political, and scientific factors that have contributed to climate change denial becoming a prevalent and enduring phenomenon.

The emergence of climate change denial can be traced back to the mid-20th century, but its roots delve deeper into history. Initially, skepticism centered on questioning whether the Earth was indeed warming. However, as scientific consensus solidified around the reality of climate change, denial shifted towards disputing its causes and implications.

Historical events have played a pivotal role in the evolution of climate change denial. The Cold War era, marked by ideological conflicts and the prominence of free-market capitalism, provided fertile ground for skepticism. It was during this period that climate change became entangled with political ideologies, with some viewing climate action as a threat to economic freedom and individual liberties.

Key figures and organizations have also left indelible marks on the landscape of climate change denial. Influential individuals, often associated with conservative think tanks and industry interests, have propagated skepticism through media outlets, books, and advocacy campaigns. These actors have played a significant role in shaping public perception and influencing policymakers.

The emergence of the internet and the proliferation of information channels have further fueled the spread of climate change denial. Skeptical voices have found platforms to disseminate their views widely, often outside the traditional realms of scientific discourse. Online communities and social media have allowed the echo chamber effect to amplify skepticism.

What If It's True?

Confronting climate change denial necessitates an exploration of the hypothetical scenarios that would unfold if the skepticism were to be validated. While the overwhelming scientific consensus supports the reality of climate change, contemplating the consequences of denial is essential for comprehending the gravity of the issue and the risks associated with inaction.

If climate change were indeed a hoax or grossly exaggerated, the ramifications would be profound. Skeptics argue that efforts to combat climate change, including policies to reduce greenhouse gas emissions, are unnecessary and economically burdensome. They contend that the world could continue with business as usual without fear of catastrophic environmental consequences.

However, delving deeper reveals a starkly different picture. The scientific consensus asserts that climate change poses substantial risks to the planet and human civilization. Rising temperatures could lead to more frequent and severe heatwaves, droughts, and wildfires. Coastal regions would face increased flooding due to rising sea levels, displacing millions of people. Ecosystems would be disrupted, affecting biodiversity and food security.

If climate change were a hoax, the world might abandon vital efforts to transition to sustainable energy sources, reduce carbon emissions, and adapt to the changing climate. This inaction could lead to irreversible damage to the environment, exacerbating existing ecological and social challenges. The costs of addressing the consequences of unchecked climate change could far exceed those of proactive mitigation and adaptation measures.

Moreover, the geopolitical implications of climate change denial are substantial. International cooperation to address climate change, exemplified by agreements like the Paris Agreement, could falter. This breakdown in cooperation might lead to heightened competition for dwindling resources, potentially exacerbating conflicts and global instability.

In exploring the hypothetical scenarios if climate change denial were true, we gain a deeper appreciation of the critical importance of addressing this issue based on scientific consensus. It underscores the need for informed decision-making, responsible policies, and global cooperation to mitigate the risks posed by climate change and secure a sustainable future for our planet.

The Scientific Foundation of Climate Change

Our quest to unravel climate change denial leads us to the bedrock of scientific consensus—the overwhelming agreement among climate scientists that the Earth's climate is warming due to human activities. In this subchapter, we delve into the comprehensive scientific foundation that underpins the reality of climate change, exploring the evidence, principles, and mechanisms that leave no room for doubt.

Central to our understanding of climate change is the meticulous analysis of temperature records. Historical temperature data, including records of global surface temperatures, reveal a consistent and undeniable trend of rising temperatures over the past century. These temperature increases align with the industrial era and the significant increase in greenhouse gas emissions from human activities.

The role of greenhouse gases, such as carbon dioxide (CO_2), methane (CH_4), and nitrous oxide (N_2O), in driving climate change is a fundamental scientific principle. These gases, emitted from sources like fossil fuel combustion and deforestation, trap heat in the Earth's atmosphere through the greenhouse effect. This natural phenomenon is well-understood and has been confirmed through laboratory experiments and observations of other planets in our solar system.

Proxy records, including ice cores, tree rings, and sediment layers, provide a window into Earth's climate history. These records offer compelling evidence of past climate variations and the role of greenhouse gases in shaping temperature patterns. The correlation between elevated CO_2 levels and temperature increases in these records reinforces our understanding of the greenhouse effect.

The scientific consensus on climate change extends to the attribution of observed warming to human activities. Climate models, based on fundamental principles of physics and chemistry, simulate temperature trends that closely match observed data when human influences are considered. This attribution is further supported by the identification of distinct fingerprints of human-induced climate change, such as the enhanced warming of the lower atmosphere and cooling of the upper atmosphere.

Dissecting Climate Change Denial

In our pursuit of understanding and addressing climate change denial, it is essential to systematically dissect and critically evaluate the central claims put forth by deniers. By subjecting these claims to scientific scrutiny and empirical evidence, we can unravel the misconceptions and inaccuracies that underpin climate change denial.

One of the primary claims of climate change denial centers on the notion that climate scientists have manipulated data to exaggerate warming trends. This assertion has been thoroughly investigated and debunked. Temperature records and climate data undergo rigorous scrutiny and verification processes, involving multiple independent organizations worldwide. The consistency of warming trends across various data sets reinforces the authenticity of temperature increases.

Another claim posited by deniers questions the reliability of climate models used to project future climate scenarios. While models inherently entail uncertainties, they have consistently demonstrated their accuracy in reproducing past climate trends and are essential tools for understanding potential future impacts of climate change. Multiple models from different research institutions converge on similar projections, bolstering their reliability.

Deniers often challenge the role of human activities in driving climate change, suggesting that natural factors alone can explain temperature variations. However, comprehensive assessments by climate scientists have unequivocally attributed recent warming to human-induced greenhouse gas emissions. The enhanced greenhouse effect, well-supported by physical principles and empirical evidence, is a fundamental component of our understanding of climate change.

Additionally, claims that climate change is merely a natural cycle are refuted by the unprecedented speed and magnitude of contemporary warming. The rate of temperature increase observed over the past century far exceeds natural variability, and the correlation between greenhouse gas emissions and warming trends is striking.

Furthermore, denial often involves cherry-picking isolated data or events to cast doubt on climate change. To counteract this tactic, comprehensive assessments consider a broad range of evidence, including temperature records, ice melt, sea-level rise, and shifts in ecosystems. The convergence of these indicators paints a compelling picture of a planet undergoing significant and rapid change.

The Role of Science Communication

In the intricate web of climate change denial, the effectiveness of science communication emerges as a pivotal factor. Communicating the complexities of climate science to the public, policymakers, and diverse audiences is essential in dispelling skepticism and fostering informed decision-making. This subchapter delves into the multifaceted landscape of science communication, examining its challenges, strategies, and the imperative of promoting accurate information.

Communicating climate science involves bridging the gap between scientific expertise and public understanding. One of the foremost challenges is the inherent complexity of the subject matter, which often requires translating intricate scientific findings into accessible language. Moreover, conveying the nuances of uncertainty while maintaining the urgency of climate action poses a delicate balancing act.

In the era of digital media and the internet, misinformation can spread rapidly, complicating science communication efforts. Climate change deniers often utilize online platforms and social media to disseminate misleading information, creating echo chambers of skepticism. Countering this requires proactive efforts to promote accurate information and fact-checking.

Effective science communication strategies encompass a range of approaches. Engaging narratives, storytelling, and visual communication can make climate science more relatable and engaging to diverse audiences. Scientists, educators, and communicators play vital roles in conveying the human and ecological dimensions of climate change.

Building public trust in science is a cornerstone of effective communication. Transparency, openness, and acknowledging uncertainties are essential components of fostering trust. Communicators must also be attuned to the values and concerns of their target audiences to effectively address their specific questions and doubts.

Furthermore, science communication extends beyond educating the public; it also influences policy decisions. Policymakers require access to clear, concise, and up-to-date scientific information to make informed choices about climate policies. Scientists must engage with policymakers and ensure that scientific findings are integrated into decision-making processes.

Conclusion

As we arrive at the culmination of our exploration into climate change denial, it is imperative to reiterate the profound significance of addressing this skepticism and embracing the science of climate change. The verdict is unequivocal: the overwhelming consensus among climate scientists underscores the reality of global warming driven by human activities. In this concluding subchapter, we recapitulate the key findings and affirm the critical importance of informed decision-making and collective action to combat climate change.

Our journey through the landscape of climate change denial has revealed a spectrum of beliefs, from outright denial to assertions of exaggeration. While skepticism can serve as a natural part of scientific inquiry, the convergence of evidence, the reliability of climate models, and the attribution of warming to human activities leave no room for doubt within the scientific community.

The potential consequences of unchecked climate change are profound and extend beyond rising temperatures. From more frequent and severe extreme weather events to threats to biodiversity and food security, the impacts of climate change are multifaceted. Addressing these challenges requires informed policies, international cooperation, and individual actions to reduce greenhouse gas emissions.

Embracing climate science is not merely an academic exercise; it is a call to action. The urgency of addressing climate change requires a collective commitment to sustainability, renewable energy, and conservation efforts. International agreements, such as the Paris Agreement, serve as vital frameworks for global cooperation in mitigating climate change.

Science communication plays a pivotal role in this endeavor, bridging the gap between scientific expertise and public understanding. Clear and engaging communication of climate science is essential for empowering individuals and policymakers to make informed decisions about climate policies and personal actions.

Finally, our journey through climate change denial highlights the undeniable reality of human-caused global warming. Climate change is a pressing global challenge, according to scientific consensus and empirical evidence. Acceptance of climate science demonstrates our commitment to a sustainable future for our planet and future generations. We can address climate change with the urgency and seriousness it requires by acknowledging the evidence, embracing informed discourse, and taking collective action.

Chapter 5: Confronting Vaccine Misinformation

In the annals of misinformation and conspiracy theories, few topics have generated as much controversy and confusion as vaccines. Welcome to the realm of "Confronting Vaccine Misinformation," a chapter dedicated to unraveling the intricate web of claims and theories that have cast doubt on one of the most potent tools in public health—the vaccine. While vaccines have undeniably saved countless lives by preventing devastating diseases, vaccine misinformation has sown seeds of skepticism, from unfounded links to autism to suspicions of government control. In this chapter, we embark on a journey to comprehensively explore the multifaceted landscape of vaccine misinformation. We will delve into the core claims that underpin vaccine skepticism, trace the historical roots of this phenomenon, contemplate the potential consequences if vaccine misinformation were validated, and reiterate the pivotal role of vaccines in safeguarding public health based on irrefutable scientific evidence.

Vaccine misinformation encompasses a range of beliefs, with some theories asserting that vaccines cause severe adverse effects, including autism, while others allege sinister motives behind vaccination campaigns. The prevalence of these beliefs has sparked controversy, stoked vaccine hesitancy, and led to declines in vaccination rates in various parts of the world. Our exploration seeks to shed light on the origins, motivations, and implications of vaccine misinformation, ultimately advocating for informed and science-based vaccination decisions as a cornerstone of public health and disease prevention. As we embark on this journey, let us examine the evidence, scrutinize the claims, and reaffirm the life-saving power of vaccines in an era marked by skepticism and misinformation.

Within the landscape of vaccine misinformation, a diverse array of claims and theories has emerged, challenging the well-established scientific consensus on the safety and efficacy of vaccines. In this subchapter, we will embark on a thorough examination of the central claims that fuel vaccine misinformation, peeling back the layers of skepticism and dissecting the myths that have gained traction among some communities.

At the forefront of vaccine misinformation stands the claim that vaccines, particularly those given during childhood, are causally linked to autism spectrum disorders (ASD). This theory gained widespread attention following the publication of a now-discredited study in the late 1990s. Despite extensive scientific research and subsequent debunking of this claim, it continues to resonate with some individuals and communities.

Beyond the autism-vaccine link, vaccine misinformation often alleges that vaccines are laden with harmful ingredients, such as mercury or thimerosal, that pose severe health risks. These assertions overlook the rigorous safety assessments and regulatory oversight governing vaccine production and distribution.

Another facet of vaccine misinformation involves conspiracy theories, suggesting that vaccines are tools for government control or covert population manipulation. These claims often imply nefarious intentions behind vaccination programs, from tracking individuals through microchips to reducing population growth.

To confront vaccine misinformation effectively, we must subject these central claims to scientific scrutiny and empirical evidence, seeking to separate fact from fiction. Our journey through this subchapter will equip us with the knowledge to address these assertions and reinforce the importance of vaccines as a cornerstone of public health and disease prevention.

To gain a deeper understanding of vaccine misinformation, we must embark on a historical journey that unveils the roots and evolution of this skepticism. The historical context surrounding vaccines is a complex tapestry, shaped by various factors, including scientific advancements, public health campaigns, and societal shifts. In this subchapter, we delve into the historical development of vaccine misinformation, exploring the events, beliefs, and circumstances that have contributed to its emergence and persistence.

The history of vaccines is intertwined with the history of infectious diseases and their devastating impact on human populations. Throughout centuries, diseases like smallpox, polio, and measles wreaked havoc, causing illness, disability, and death. The development of vaccines marked a monumental breakthrough in public health, offering protection against these deadly scourges.

The success of vaccination programs in reducing the prevalence of once-deadly diseases led to a sense of security and complacency. As generations grew up without witnessing the devastating consequences of diseases like smallpox, concerns about vaccine safety began to emerge. These concerns laid the groundwork for skepticism.

The origins of vaccine misinformation can be traced back to isolated incidents and case reports that fueled fears about vaccine safety. The infamous Wakefield study, published in 1998 and later discredited, falsely claimed a link between the MMR vaccine and autism, sparking widespread concerns and controversy.

The rise of the internet and social media platforms provided a new avenue for the spread of vaccine misinformation. Online communities and websites propagated unfounded claims, amplifying vaccine skepticism and undermining public trust in vaccination.

What If It's True?

In our quest to address vaccine misinformation, it is essential to confront the hypothetical scenarios that would unfold if the claims of skeptics were validated. While the overwhelming scientific consensus supports the safety and effectiveness of vaccines, contemplating the consequences of vaccine misinformation becoming accepted is essential for comprehending the gravity of the issue and the risks associated with vaccine hesitancy and refusal.

If the claim that vaccines, particularly childhood vaccines, were indeed linked to autism or other severe health issues were validated, the implications would be profound. Parents and caregivers would understandably hesitate to vaccinate their children, fearing the potential risks. This hesitation could lead to a decline in vaccination rates, resulting in a resurgence of vaccine-preventable diseases that were once under control.

The consequences of reduced vaccination rates extend beyond individual health. Communities and populations would become vulnerable to outbreaks of infectious diseases. Diseases like measles, mumps, and rubella could regain a foothold, posing risks to both children and adults. Healthcare systems would be strained as they grapple with preventable disease outbreaks.

Furthermore, the erosion of trust in vaccines could have a cascading effect, undermining confidence in other public health measures and recommendations. The ability to respond effectively to future health crises, such as pandemics, could be compromised if vaccine hesitancy and misinformation persist.

The Scientific Foundation of Vaccination

To counter vaccine misinformation effectively, it is crucial to delve into the scientific foundation that underpins vaccination as a cornerstone of public health. In this subchapter, we will explore the comprehensive scientific consensus on vaccination, presenting empirical evidence on vaccine safety and efficacy, and discussing the role of vaccines in preventing diseases and ensuring public health.

At the heart of vaccination lies the concept of immunization, a process by which the immune system is primed to recognize and defend against specific pathogens. Vaccines are designed to mimic the presence of a pathogen, stimulating the immune response without causing the disease itself. This process equips the body with a memory of the pathogen, enabling it to mount a rapid and effective defense if exposed to the real pathogen in the future.

The safety of vaccines is rigorously assessed through extensive clinical trials and post-marketing surveillance. Vaccine development and approval involve rigorous testing to ensure their safety and efficacy. Regulatory agencies worldwide, such as the U.S. Food and Drug Administration (FDA) and the European Medicines Agency (EMA), oversee the evaluation and approval of vaccines.

Empirical evidence overwhelmingly supports the safety and efficacy of vaccines. Vaccination has successfully eradicated or greatly reduced the incidence of numerous diseases, including smallpox and polio. Diseases like measles, which once caused significant morbidity and mortality, have been brought under control through vaccination efforts.

Vaccines have not only saved lives but have also reduced healthcare costs and prevented long-term disabilities. The economic benefits of vaccination extend beyond healthcare, contributing to overall societal well-being and productivity.

Dissecting Vaccine Misinformation

To effectively counter vaccine misinformation, we must systematically dissect the central claims made by proponents of skepticism and critically evaluate the evidence they present. In this subchapter, we embark on a journey through the maze of vaccine myths, addressing common misconceptions, misinterpretations, and inaccuracies that have fueled vaccine hesitancy and refusal.

At the forefront of vaccine misinformation stands the claim that vaccines, particularly those administered during childhood, are causally linked to autism spectrum disorders (ASD). This belief traces its origins to a study published in 1998 by Dr. Andrew Wakefield, which suggested a connection between the MMR (measles, mumps, and rubella) vaccine and autism. However, this study was later discredited, and its author faced professional repercussions, including the retraction of the paper. Extensive subsequent research, involving millions of individuals, has consistently failed to establish any such link between vaccines and autism. The scientific consensus is resolute: vaccines do not cause autism.

Another common claim posited by vaccine skeptics is that vaccines contain harmful ingredients, such as mercury or thimerosal, which pose severe health risks. These assertions overlook the rigorous safety assessments and regulatory oversight governing vaccine production and distribution. Thimerosal, a preservative used in some vaccines, has been extensively studied, and the overwhelming evidence supports its safety. In fact, thimerosal has been removed from most childhood vaccines as a precautionary measure, despite the absence of scientific evidence linking it to harm.

Conspiracy theories also play a role in vaccine misinformation, suggesting that vaccines are tools for government control or covert population manipulation. These claims often imply nefarious intentions behind vaccination programs, from tracking individuals through microchips to reducing population growth. Such assertions lack empirical evidence and are inconsistent with the transparent and collaborative nature of vaccine development and public health efforts.

To counter vaccine misinformation effectively, we must subject these central claims to scientific scrutiny, rigorous research, and empirical evidence. The overwhelming body of scientific research, the consensus among medical professionals, and the rigorous regulatory processes governing vaccines collectively affirm their safety, efficacy, and critical role in disease prevention. By dissecting vaccine misinformation and embracing the scientific foundation of vaccination, we equip ourselves with the knowledge needed to combat skepticism and promote informed vaccination decisions, safeguarding individual and public health.

The Role of Public Health and Immunization

Vaccination serves as a linchpin of public health, contributing significantly to the prevention and control of infectious diseases. In this subchapter, we explore the multifaceted role of vaccines in safeguarding public health, discussing immunization programs, disease eradication efforts, and the critical concept of herd immunity.

Immunization programs form the cornerstone of public health strategies to combat infectious diseases. These programs are meticulously designed to provide individuals and communities with protection against a wide array of preventable diseases, from childhood illnesses like measles and chickenpox to deadly viruses like influenza and hepatitis. Vaccination schedules, recommended by healthcare authorities like the Centers for Disease Control and Prevention (CDC) and the World Health Organization (WHO), outline the timing and doses necessary to build immunity.

Disease eradication efforts demonstrate the profound impact of vaccines on public health. Smallpox, one of the deadliest diseases in human history, was declared eradicated in 1980 due to a global vaccination campaign. Polio is another disease on the brink of eradication, with vaccination campaigns targeting its complete elimination.

Herd immunity, a critical concept in public health, underscores the importance of widespread vaccination. When a sufficient percentage of a population is immune to a disease, either through vaccination or previous infection, it becomes challenging for the disease to spread. Vulnerable individuals who cannot be vaccinated due to medical reasons or age, such as infants or those with compromised immune systems, are protected by the immunity of the broader community.

However, declining vaccination rates, driven in part by vaccine misinformation, can jeopardize herd immunity. This puts vulnerable populations at risk and can lead to disease outbreaks. Recent measles outbreaks in various parts of the world serve as poignant examples of the consequences of declining vaccination rates.

Conclusion

In the culmination of our exploration into vaccine misinformation, we arrive at a juncture where the significance of informed vaccination decisions and the critical role of vaccines in public health cannot be overstated. The journey through this chapter has led us through the labyrinth of skepticism and misinformation, presenting a comprehensive picture of the challenges and consequences associated with vaccine hesitancy.

Our examination of vaccine misinformation has revealed a spectrum of beliefs, from unfounded claims of vaccine-autism links to allegations of sinister government agendas. These beliefs have the potential to erode public trust in vaccines, contributing to declining vaccination rates and outbreaks of preventable diseases.

The scientific consensus on vaccination is unwavering: vaccines are safe, effective, and critical for public health. Rigorous research, extensive clinical trials, and post-marketing surveillance underpin the safety and efficacy of vaccines. Vaccination programs have successfully prevented numerous diseases and saved countless lives.

The consequences of vaccine misinformation are far-reaching, encompassing the risk of disease resurgence, strained healthcare systems, and eroded public trust in public health measures. Contemplating the hypothetical scenarios if vaccine skepticism were validated underscores the gravity of the issue and the urgency of addressing it.

As we conclude this chapter, it is essential to reaffirm the paramount importance of championing vaccination. Informed vaccination decisions, based on accurate information and scientific evidence, are an essential responsibility for individuals and communities alike. By embracing vaccination as a critical tool in disease prevention and public health, we safeguard the well-being of ourselves, our communities, and future generations. The battle against vaccine misinformation is won through knowledge, informed discourse, and a collective commitment to the health and well-being of all.

Chapter 6 Alien Cover-Ups - Unveiling Extraterrestrial Mysteries

As we journey deeper into the captivating realm of conspiracy theories, our next destination carries us beyond the boundaries of Earth itself—welcome to the chapter titled "Alien Cover-Ups: Unveiling Extraterrestrial Mysteries." In this chapter, we delve into one of the most alluring and enduring beliefs in the annals of conspiracy theories—the notion that governments worldwide are engaged in an elaborate game of cosmic hide-and-seek, concealing evidence of encounters with beings from beyond our planet. It's a narrative that has captured the imagination of many, akin to a cosmic puzzle where secrecy shrouds alleged extraterrestrial visitors. Our exploration will take us through a comprehensive analysis of the central claims surrounding these alleged cover-ups, illuminate the historical context in which these beliefs have evolved, consider the profound consequences that would unfold if such claims were validated, and, above all, reaffirm our unwavering commitment to the pursuit of knowledge grounded in the principles of scientific evidence.

At the heart of the concept of "Alien Cover-Ups" lies the compelling belief that governments, intelligence agencies, and shadowy organizations worldwide are engaged in an intricate and long-running conspiracy. This conspiracy centers around the suppression of evidence related to unidentified flying objects (UFOs) and purported encounters with extraterrestrial beings. Proponents of these claims cite government secrecy, declassified documents, and eyewitness testimonies as compelling evidence that a cosmic truth is being concealed from the public.

The historical development of beliefs in alien cover-ups is a fascinating journey through the 20th century and beyond. Early UFO sightings, including the infamous Roswell incident of 1947, played a pivotal role in shaping public perceptions and popular culture. Government initiatives, such as Project Blue Book, were launched to investigate UFO reports. These events, along with the influence of science fiction and media, laid the groundwork for enduring beliefs in clandestine extraterrestrial encounters.

However, the question remains: What if these claims of alien cover-ups were true? The potential consequences are profound, encompassing societal upheaval, scientific revolution, and geopolitical shifts. The validation of such claims would fundamentally alter our understanding of our place in the cosmos, raising questions about the existence of advanced civilizations, the nature of interstellar communication, and the implications for our own future.

As we embark on this exploration of "Alien Cover-Ups," we will systematically dissect the central claims made by proponents of these beliefs, evaluating the evidence and considering alternative explanations. We will delve into the role of science and skepticism in approaching extraordinary claims and emphasize the importance of evidence-based inquiry. Ultimately, our journey will lead us to a resounding affirmation of the value of open inquiry, critical thinking, and the pursuit of knowledge, as we seek to unveil the cosmic mysteries that continue to captivate our imagination and curiosity.

As we step into the intricate world of alien cover-ups, our first task is to scrutinize the central claims that form the bedrock of this enduring conspiracy theory. The belief that governments worldwide are hiding evidence of extraterrestrial encounters and concealing the truth about unidentified flying objects (UFOs) has captured the imagination of many. In this subchapter, we delve into the heart of these claims, examining the elements that proponents consider as compelling evidence for a cosmic conspiracy.

At the core of the concept of alien cover-ups is the assertion that governments and intelligence agencies are involved in a vast and secretive operation to suppress information about UFO sightings and encounters with extraterrestrial beings. Proponents of these claims often point to declassified government documents, whistleblower testimonies, and alleged insider leaks as key pieces of evidence supporting this belief.

Decades of purported UFO sightings, some of which defy conventional explanations, have contributed to the conviction that governments are aware of, and actively concealing, the existence of unidentified aerial phenomena. The famous Roswell incident of 1947, where an unidentified object crashed in New Mexico, has become emblematic of these beliefs.

While the U.S. military initially stated that it was a "flying disc," the official explanation later changed to a weather balloon. This shift in explanation has fueled suspicions of a government cover-up.

Eyewitness testimonies from military personnel, pilots, and civilians who claim to have encountered UFOs or witnessed unexplained aerial phenomena are often cited as compelling evidence. These accounts range from sightings of unusual aircraft to reports of encounters with beings from other worlds.

To fully appreciate the beliefs and claims surrounding alien cover-ups, we must embark on a historical journey that traces the roots and evolution of this captivating conspiracy theory. The history of UFO sightings, government investigations, and the popularization of extraterrestrial encounters forms a complex tapestry that has contributed to the enduring allure of alien cover-ups.

The phenomenon of UFO sightings, often associated with unidentified flying objects, has a long and diverse history. Reports of strange and unexplained aerial phenomena date back centuries, with accounts varying across cultures and civilizations. However, it was in the mid-20th century, particularly after World War II, that UFO sightings began to capture public attention on a global scale.

The 1947 Roswell incident, which involved the alleged crash of an unidentified object in New Mexico, played a pivotal role in shaping the modern UFO narrative. Initially described by the U.S. military as a "flying disc," the official explanation later shifted to a weather balloon. This change in explanation fueled suspicions of a government cover-up, making Roswell a focal point for conspiracy theorists.

Government initiatives aimed at investigating UFO sightings further contributed to the historical context of alien cover-ups. Project Blue Book, a U.S. Air Force program launched in the 1950s, sought to evaluate UFO reports and determine if they posed any threats to national security. While most cases were explained as natural phenomena or conventional aircraft, a small percentage remained unexplained, providing fodder for conspiracy theories.

The influence of popular culture, science fiction literature, and media also played a significant role in shaping beliefs surrounding extraterrestrial encounters. UFO sightings and the notion of government cover-ups found their way into books, movies, and television shows, further fueling public fascination with the topic.

Prominent figures and organizations, such as ufologists and UFO research groups, emerged as advocates for the investigation of UFOs and the disclosure of government secrets. These individuals and organizations contributed to the dissemination of conspiracy theories surrounding alien cover-ups.

Our exploration into the realm of alien cover-ups takes a thought-provoking turn as we consider the hypothetical scenarios that would unfold if the claims of conspiracy theorists were validated. While it's essential to approach these claims with skepticism and critical scrutiny, contemplating the consequences of their veracity is crucial for comprehending the gravity of the issue and the potential implications for society, science, and geopolitics.

If the claims of alien cover-ups were validated, the ramifications would be profound and far-reaching. Society would grapple with the revelation that we are not alone in the universe, and extraterrestrial beings have visited our planet. This paradigm shift would raise profound questions about the nature of these beings, their intentions, and the implications for human civilization.

The scientific community would face a seismic shift as well. The confirmation of extraterrestrial life would challenge our understanding of biology, evolution, and the conditions necessary for life to exist. Scientists would rush to study and learn from these otherworldly visitors, seeking insights into advanced technologies and scientific knowledge that could revolutionize our understanding of the cosmos.

On a geopolitical level, the confirmation of alien cover-ups would alter the balance of power and diplomacy on Earth. Nations would grapple with the implications of contact with extraterrestrial civilizations, potentially leading to new alliances, conflicts, or efforts at global cooperation.

Questions of how to respond to potential threats or opportunities posed by extraterrestrial beings would dominate international discussions.

Societal and cultural shifts would also be inevitable. Belief systems, religious interpretations, and cultural narratives would evolve in response to the confirmation of extraterrestrial life. Humanity would be forced to reconcile its place in the universe with the newfound knowledge of other intelligent civilizations.

While these hypothetical scenarios are captivating to consider, it is crucial to emphasize that claims of alien cover-ups lack robust scientific evidence and remain firmly in the realm of conspiracy theories. Skepticism, critical thinking, and the scientific method serve as essential tools for evaluating extraordinary claims and differentiating between evidence-based knowledge and speculation. As we navigate the cosmic puzzle of alien cover-ups, we must remain grounded in the pursuit of truth and the rigorous standards of empirical inquiry.

The Search for Extraterrestrial Life

To better contextualize the claims and beliefs surrounding alien cover-ups, we must embark on a scientific journey that explores humanity's quest for extraterrestrial life. While conspiracy theories may tantalize with their tales of hidden encounters, the scientific pursuit of understanding life beyond Earth is grounded in rigorous inquiry and empirical exploration.

At the heart of this quest is the recognition that the universe is vast and teeming with countless stars, planets, and galaxies. The sheer scale of the cosmos, with its billions of potentially habitable planets, has fueled scientific curiosity about the potential for life beyond our own world.

One of the foundational concepts in this exploration is the Drake Equation, a mathematical formula designed to estimate the number of technologically advanced civilizations in our galaxy with which we might potentially communicate. While the equation involves numerous variables, including the rate of star formation and the fraction of planets capable of supporting life, its application underscores the plausibility of extraterrestrial civilizations.

Conversely, the Fermi Paradox poses a thought-provoking question: If extraterrestrial civilizations are probable, why haven't we detected any signs of them? This paradox highlights the challenges of interstellar communication and the vast distances that separate stars and galaxies.

Scientific efforts to search for extraterrestrial life extend to the study of exoplanets, celestial bodies orbiting stars outside our solar system. Discoveries of thousands of exoplanets, some within the habitable zone where liquid water could exist, have further fueled speculation about the potential for life elsewhere.

While the search for extraterrestrial life remains ongoing, it is essential to emphasize that the scientific approach is grounded in evidence, observation, and the rigorous application of the scientific method. Claims of alien cover-ups often lack these critical foundations, relying instead on anecdotal accounts and speculation.

As we navigate the intricate tapestry of beliefs surrounding alien cover-ups, it is essential to differentiate between the scientific pursuit of understanding life in the cosmos and the speculative narratives of conspiracy theories. Our commitment to knowledge rests on the principles of evidence-based inquiry and empirical exploration, as we continue to explore the mysteries of the universe with curiosity and rigor.

Dissecting Alien Cover-Ups

In our quest to unravel the enigmatic world of alien cover-ups, we undertake a meticulous examination of the central claims that underpin this intriguing conspiracy theory. These claims are often presented as compelling evidence of clandestine extraterrestrial encounters and government secrecy. However, it is through critical analysis and scientific scrutiny that we endeavor to distinguish between the speculative and the substantiated.

At the forefront of these claims lies the notion of government secrecy, purportedly evidenced by declassified documents related to unidentified flying objects (UFOs) and extraterrestrial encounters. Proponents assert that the release of such documents implies a hidden truth. While it is essential to acknowledge the existence of declassified materials, we must consider the context in which they were made public. Declassification often occurs for reasons unrelated to alien cover-ups, such as national security concerns or the passage of time. Rigorous evaluation of these documents necessitates a thorough understanding of their content, origin, and significance within the broader context of government operations.

Eyewitness testimonies, another pillar of alien cover-up claims, offer personal accounts of encounters with UFOs and extraterrestrial beings. These testimonies are deeply compelling, as they provide firsthand narratives of unexplained phenomena. However, it is crucial to recognize the limitations of human perception and the potential for cognitive biases. Factors such as misidentifications of natural or man-made objects, optical illusions, and the influence of popular culture can shape eyewitness accounts. Thus, while these testimonies may be sincerely held, they require critical scrutiny and corroborating evidence to establish their credibility.

Whistleblower testimonies, often touted as insider revelations of government involvement in alien cover-ups, introduce a complex layer of claims. These individuals claim to possess privileged information regarding secret programs and concealed truths. Evaluating the veracity of whistleblower claims involves a comprehensive assessment of their motives, credibility, and the evidence they present. Careful consideration is also given to the presence of corroborating witnesses or documentation to support their allegations.

Misidentifications of natural or conventional phenomena play a significant role in UFO sightings and encounters. Weather phenomena, astronomical events, and human-made objects such as aircraft and satellites can all contribute to sightings that defy immediate explanation. In many instances, thorough investigation reveals prosaic and scientifically sound explanations, dispelling the notion of extraterrestrial involvement.

Cognitive biases and the innate human desire for mystery further shape beliefs in alien cover-ups. The confirmation bias, which inclines individuals to seek and interpret information that confirms their preexisting beliefs, can reinforce conspiracy narratives. The allure of the unexplained, coupled with a tendency to attribute unknown phenomena to extraordinary causes, can contribute to the perception of cover-ups.

In dissecting the central claims of alien cover-ups, we emphasize the scientific approach to extraordinary claims. Science relies on empirical evidence, peer review, and transparent methodologies to evaluate assertions rigorously. The pursuit of credible knowledge necessitates a commitment to objective analysis, critical thinking, and the rigorous examination of evidence. As we navigate the intricate web of beliefs surrounding alien cover-ups, our dedication to evidence-based inquiry remains unwavering, guiding us toward a deeper understanding of the mysteries that continue to captivate our imagination.

The Role of Science and Skepticism

In our continued journey through the intriguing world of alien cover-ups, we turn our attention to the pivotal role of science and skepticism in addressing extraordinary claims and navigating the complex landscape of conspiracy theories. The scientific method, anchored in empirical evidence and systematic inquiry, offers a powerful tool for evaluating assertions that transcend conventional knowledge.

1. **The Scientific Method:** At the heart of the scientific approach lies the systematic process of inquiry known as the scientific method. It involves formulating hypotheses, conducting experiments, gathering data, and subjecting findings to rigorous analysis. Science relies on empirical evidence, reproducibility, and the pursuit of objective truth. It provides a structured framework for distinguishing between credible knowledge and unfounded speculation.

2. **Critical Thinking and Skepticism:** Skepticism, characterized by a healthy questioning of claims and a demand for evidence, serves as a cornerstone of scientific inquiry. Critical thinking, coupled with skepticism, encourages individuals to assess assertions with impartiality and rigor. It allows us to navigate the sea of information and misinformation, fostering a discerning approach to extraordinary claims.

3. **Extraordinary Claims Require Extraordinary Evidence:** The scientific community adheres to the principle that extraordinary claims demand extraordinary evidence. When evaluating

assertions as profound as extraterrestrial encounters and government cover-ups, a higher standard of evidence is essential. Robust, peer-reviewed research, independently verified data, and the absence of alternative explanations strengthen the credibility of claims.

4. **Transparency and Open Investigation:** Scientific inquiry thrives on transparency and open investigation. Findings are subject to peer review, where experts in the field critically assess research methodologies and conclusions. Open discourse and the willingness to reevaluate findings based on new evidence are fundamental aspects of the scientific process.

5. **Balancing Curiosity and Rigor:** While curiosity fuels our exploration of the unknown, it must be balanced with intellectual rigor. Scientific inquiry is driven by the quest for truth, not the reinforcement of preconceived notions. Open-mindedness, combined with rigorous analysis, fosters an environment where objective knowledge can flourish.

In the context of alien cover-ups, the principles of science and skepticism serve as powerful tools for discerning between credible knowledge and speculation. The commitment to empirical evidence, critical thinking, and the pursuit of objective truth guides us in our exploration of cosmic mysteries. As we navigate the intricate terrain of conspiracy theories, the light of scientific inquiry illuminates our path, encouraging us to embrace knowledge grounded in evidence and to question the shadows of unfounded belief.

Conclusion

Our odyssey through the realm of alien cover-ups reaches its zenith as we conclude our exploration of this beguiling conspiracy theory. This chapter has been a voyage through the captivating narratives of hidden encounters, government secrecy, and the tantalizing prospect of extraterrestrial life. As we prepare to embark on new intellectual adventures, we pause to reflect on the significance of our journey.

The examination of alien cover-ups has unveiled a complex landscape of beliefs, claims, and narratives that have captured the human imagination for decades. From government secrecy and declassified documents to eyewitness testimonies and whistleblower revelations, we have traversed a diverse terrain of assertions. Yet, it is through the lens of science and skepticism that we have sought clarity amidst the enigma.

While the allure of cosmic mysteries and government cover-ups may persist, we are reminded of the steadfast commitment to evidence-based inquiry and the pursuit of objective knowledge. The principles of the scientific method, critical thinking, and skepticism have served as guiding stars, illuminating our path through the cosmic labyrinth.

As we conclude our exploration of alien cover-ups, we reiterate the importance of embracing knowledge grounded in empirical evidence and of questioning assertions that transcend conventional understanding. Our journey through the intricate tapestry of conspiracy theories has reinforced the value of discernment and intellectual rigor.

Chapter 7: HAARP Weather Control - Unraveling the Conspiracy

Welcome, intrepid explorers, to the enigmatic realm of the High-Frequency Active Auroral Research Program, or HAARP, and the shadowy conspiracy theories that shroud it in mystery. In this chapter, we embark on a voyage into the intricate world of HAARP, an advanced scientific facility, and the extraordinary claims that have surrounded it. Some assert that HAARP possesses the power to control weather patterns, while others go even further, suggesting mind-controlling capabilities. Together, we will unravel these intriguing assertions and explore the truth behind HAARP.

HAARP, situated in the remote wilderness of Alaska, is a research program designed to study the Earth's ionosphere—a region of the atmosphere charged with charged particles and high-frequency radio waves. While its primary mission is the pursuit of scientific knowledge, HAARP has found itself entangled in a web of conspiracy theories, captivating the imaginations of those who seek to decipher its mysteries.

In this chapter, we will embark on a multifaceted journey. We will begin by examining the central claims that have made HAARP a focal point of speculation and intrigue. These claims encompass weather manipulation on a global scale and even the ability to control human minds. As we delve deeper into these assertions, we will scrutinize the evidence and narratives presented by proponents of these conspiracy theories.

Our exploration will then take us back in time, tracing the origins and functions of HAARP. We will uncover the scientific objectives that underpin this facility and explore the technology behind its high-frequency radio waves. Government involvement and potential military applications will also be unveiled to provide a comprehensive historical context.

As we contemplate the ramifications of these claims, we will ponder the profound "what if." What if the assertions of HAARP's weather control and mind control capabilities were validated? What ethical dilemmas and environmental consequences might arise? How would our understanding of technology's capabilities be reshaped?

However, we must not lose sight of the rigorous principles of scientific inquiry and skepticism. In our relentless pursuit of truth, we will systematically debunk the central claims made by proponents of HAARP conspiracy theories. Scientific principles, technological limitations, and the scientific consensus on HAARP's actual capabilities will be our guiding lights in this endeavor.

Throughout our journey, we will underscore the pivotal role of science and skepticism. The scientific method, characterized by empirical evidence and rigorous analysis, serves as our compass through the maze of conjecture. Skepticism and critical thinking will illuminate our path, guiding us toward objective knowledge.

As we venture deeper into the heart of the HAARP conspiracy theories, we remain committed to the pursuit of truth grounded in evidence. The enigmatic world of HAARP awaits our exploration, beckoning us to unravel its secrets and embrace the guiding light of scientific inquiry.

Our journey into the enigmatic realm of HAARP commences with the central claims that have given rise to elaborate conspiracy theories. These claims, though extraordinary, form the cornerstone of the intrigue surrounding the High-Frequency Active Auroral Research Program.

At the forefront of these assertions is the belief that HAARP possesses the formidable ability to manipulate weather patterns on a global scale. Proponents contend that by directing its high-frequency radio waves toward the Earth's ionosphere, HAARP can generate extreme weather events such as hurricanes, tornadoes, or droughts. Throughout this subchapter, we will meticulously scrutinize the evidence and narratives presented to substantiate this remarkable claim.

Taking the conspiracy theories a step further, some allege that HAARP's reach extends into the realm of human cognition and behavior. The notion of mind control suggests that HAARP can transmit signals capable of influencing human brains, potentially leading to alterations in thought processes, moods, or behaviors. As we explore this claim, we will delve into the specifics and evaluate the supporting evidence.

Intriguingly, these conspiracy theories also raise concerns about the environmental impact of HAARP's activities. The manipulation of the ionosphere, as posited by the claims, could have profound ecological consequences, including disruptions in climate patterns and disturbances in the Earth's magnetic field. We will delve into the potential ramifications and consider the scientific plausibility of such assertions.

Central to these claims is the concept of government secrecy and covert operations. Proponents argue that the true extent of HAARP's capabilities is concealed behind a veil of classified experiments and clandestine activities. We will critically assess the evidence and arguments put forth to support this notion.

As we embark on this exploratory journey, we cannot overlook the global implications of these claims. If allegations of HAARP's weather manipulation and mind control were substantiated, how might other nations react? What geopolitical reverberations might ensue from the validation of these conspiracy theories? These are questions we will ponder as we navigate the intricate web of assertions, guided by the principles of critical thinking, scientific inquiry, and the relentless pursuit of evidence-based knowledge.

To unravel the enigma of the High-Frequency Active Auroral Research Program (HAARP) and the conspiracy theories surrounding it, we must embark on a historical journey that sheds light on HAARP's origins, functions, and the scientific objectives that underpin its existence. By delving into HAARP's backstory, we can gain a deeper understanding of the facility and its intended purposes.

HAARP's origins can be traced back to the late 20th century when the U.S. government, in collaboration with the U.S. Navy and the Defense Advanced Research Projects Agency (DARPA), initiated the program. The primary aim was to conduct research in the Earth's ionosphere, a region of the atmosphere characterized by charged particles and high-frequency radio waves. The ionosphere plays a crucial role in global communication and navigation systems, making it an area of significant scientific interest.

At the core of HAARP's research is the utilization of high-frequency radio waves to stimulate and probe the ionosphere. This technology enables scientists to gain insights into the ionosphere's properties and behavior, enhancing our understanding of its impact on radio communications, satellite operations, and global positioning systems.

HAARP's scientific endeavors have included investigations into phenomena such as ionospheric disturbances, radio wave propagation, and the aurora borealis, shedding light on natural processes that have fascinated scientists for generations. Additionally, the facility has served as a platform for international collaborations, fostering advancements in ionospheric research.

Government involvement in HAARP's funding and operation is a crucial aspect of its history. While the facility's research objectives are rooted in scientific inquiry, it has also drawn interest from military and defense sectors. This dual-purpose nature has contributed to speculation and conspiracy theories, as some believe that HAARP's capabilities extend beyond its stated scientific objectives.

What If It's True?

In our exploration of the conspiracy theories enveloping the High-Frequency Active Auroral Research Program (HAARP), we find ourselves contemplating the profound hypothetical scenarios that would unfold if these claims of weather manipulation and mind control were indeed validated.

Firstly, envision a world where HAARP's asserted ability to manipulate weather patterns globally becomes a reality. The consequences of such power would be nothing short of staggering. Extreme weather events, whether they be hurricanes, droughts, or tornadoes, could be generated at will, posing substantial risks to agriculture, infrastructure, and human populations. The potential to disrupt ecosystems and economies raises critical questions about the ethical and environmental implications of wielding such extraordinary capabilities.

Secondly, consider the implications if HAARP's alleged mind control capabilities were unveiled as fact. This scenario delves into uncharted territory, challenging our understanding of technology's influence on human cognition and behavior. Fundamental questions regarding consent, individual autonomy, and the potential for misuse would come to the forefront of public discourse. Concerns about safeguarding personal freedoms and privacy would necessitate profound societal reflections.

Moreover, our exploration extends to the broader environmental consequences that might arise from tampering with the ionosphere, as suggested by these conspiracy theories. Speculations regarding disturbances in climate patterns and disruptions in the Earth's magnetic field compel us to contemplate the intricate interconnectedness of our planet's ecosystems and the fragility of natural processes.

Furthermore, if the claims of government secrecy surrounding HAARP were substantiated, profound questions about transparency, accountability, and the boundaries of classified operations would emerge. The revelation of concealed agendas and covert experiments would cast a spotlight on the trustworthiness of government institutions and the democratic foundations upon which they are built.

Lastly, the global repercussions of validating HAARP conspiracy claims are a topic of paramount importance. The international community would face the challenge of adapting to a world where weather manipulation and mind control capabilities exist. Diplomatic relations, security considerations, and global governance structures could undergo seismic shifts as nations grapple with the implications of these newfound realities.

Debunking HAARP Conspiracy Claims

As our journey through the labyrinth of HAARP conspiracy theories continues, we turn our focus to the critical task of systematically debunking the central claims that have given rise to intrigue and speculation. To distinguish between conjecture and credible information, we must subject these claims to rigorous scrutiny and evaluate them in the context of scientific principles, technological limitations, and empirical evidence.

Firstly, let us address the claim of weather manipulation attributed to HAARP. While the notion of controlling weather patterns on a global scale is tantalizing, it stands at odds with the scientific understanding of meteorology and atmospheric processes. Weather systems are extraordinarily complex, influenced by a multitude of variables, and subject to natural laws that are beyond the scope of HAARP's technology.

Scientific consensus aligns with the view that HAARP's capabilities do not extend to the manipulation of weather phenomena.

Secondly, the assertion of mind control attributed to HAARP demands careful consideration. The human brain is a vastly intricate organ, and the idea that high-frequency radio waves could exert influence over it raises profound questions about neuroscience and the limitations of technology. Empirical studies and neurological research offer no substantiated evidence to support claims of HAARP-induced mind control. The scientific consensus asserts the implausibility of such capabilities.

Moreover, when evaluating claims of environmental consequences arising from HAARP's activities, we must turn to the principles of environmental science and ecology. The notion that HAARP's operations could trigger disturbances in climate patterns or disrupt the Earth's magnetic field is not supported by scientific research. Our understanding of environmental systems indicates that these assertions lack empirical foundation.

Furthermore, the claim of government secrecy and covert operations at HAARP is a complex matter. While governments do engage in classified research, the leap from classified activities to global weather manipulation and mind control remains unsubstantiated. The absence of verifiable evidence supporting these claims and the adherence to scientific protocols underscore the importance of skepticism and critical inquiry.

The Role of Science and Skepticism

In our ongoing exploration of the HAARP conspiracy theories, it is imperative to pause and reflect on the pivotal role that science and skepticism play in our quest for understanding. These two principles are our guiding lights as we navigate the complex landscape of extraordinary claims and seek to distinguish fact from fiction.

Firstly, the scientific method stands as an unwavering pillar of evidence-based inquiry. Rooted in empirical evidence and rigorous analysis, it serves as our compass through the maze of conjecture. Science demands meticulous observation, experimentation, and peer review to validate hypotheses and theories. When applied to the claims surrounding HAARP, this method highlights the need for credible evidence and sound scientific reasoning to support assertions of weather manipulation and mind control.

Secondly, skepticism emerges as a critical tool in our arsenal. The discerning eye of skepticism prompts us to question, scrutinize, and challenge assertions that transcend conventional understanding. It encourages us to assess claims critically and to consider alternative explanations. Skepticism serves as a safeguard against the embrace of unfounded beliefs and invites us to seek robust evidence before accepting extraordinary claims.

Furthermore, critical thinking is our ally in this endeavor. It empowers us to analyze information objectively, evaluate sources of information, and assess the validity of arguments. Critical thinking invites us to approach claims with a discerning mindset, to distinguish between reliable and unreliable sources, and to recognize logical fallacies or biases that may cloud our judgment.

Conclusion

Our odyssey through the realm of HAARP and its associated conspiracy theories reaches its culmination as we arrive at the conclusion of our exploration. This chapter has been a voyage into the intricate web of claims, counterclaims, and enigma that have surrounded the High-Frequency Active Auroral Research Program. As we prepare to draw our journey to a close, we reflect on the significance of our quest.

The examination of HAARP conspiracy theories has unveiled a complex tapestry of beliefs and assertions that have captured the human imagination. From weather manipulation to mind control, from government secrecy to environmental consequences, we have traversed a diverse landscape of claims. Yet, it is through the lens of science, skepticism, and critical thinking that we have sought clarity amidst the complexity.

While the temptation of remarkable claims may endure, we are reminded of the need of evidence-based inquiry and the search of objective knowledge. The scientific method, skepticism, and critical thinking have served as guiding stars, illuminating our path through the maze of conspiracy theories.

As we conclude our investigation of HAARP and its accompanying mysteries, we reiterate the importance of accepting empirically based information and questioning assertions that defy common thinking. Our journey through the complicated tapestry of conspiracy theories has highlighted the importance of judgment and intellectual rigor.

Chapter 8: Hollow Earth Theory

Welcome, intrepid explorers, to a chapter that delves into the depths of an age-old mystery—the Hollow Earth Theory. Imagine, if you will, a world beneath our feet, not of molten rock and solid earth, but a vast, hollow realm that may conceal secrets beyond our wildest imaginings. This captivating hypothesis, akin to the fantastical journeys depicted in Jules Verne's "Journey to the Center of the Earth," has captivated the human imagination for centuries. It beckons us to venture beyond the boundaries of conventional knowledge, inviting us to contemplate the possibility of a hidden world, complete with its own landscapes, ecosystems, and perhaps even advanced civilizations.

The Hollow Earth Theory stands as a testament to the enduring allure of the unknown. It entices us to envision subterranean realms where the laws of science and the boundaries of reality blur, allowing our imaginations to run wild. Yet, this theory is not merely a product of fiction or whimsy; it has a rich history, with roots that extend deep into the annals of human thought. From ancient civilizations to contemporary explorers of the mind, the concept of a hollow Earth has manifested in various forms, adapting and evolving with the passage of time.

As we embark on this exploration, we will journey through the layers of the Hollow Earth Theory, peeling back the layers of speculation to uncover its historical origins, the imaginative narratives it has inspired, and the scientific refutations it has faced. Through the lens of critical inquiry and empirical evidence, we will attempt to discern the truth behind the claims and gain a deeper understanding of the enduring appeal of this captivating enigma. So, join us as we venture into the mysterious depths of the Hollow Earth Theory, where fact and fiction converge in a quest for enlightenment and discovery.

To begin our journey into the Hollow Earth Theory, we must first grapple with its central claims—claims that propose a reality vastly different from the one we perceive on the surface. At the heart of this theory lies the extraordinary assertion that our Earth is not a solid sphere but rather a hollow structure, with a vast, uncharted inner world. Within this hollow realm, proponents of the theory suggest the existence of an inner sun, providing light, heat, and sustenance to an advanced civilization dwelling beneath the Earth's crust.

This audacious theory further postulates that this subterranean civilization is far more advanced than our own, possessing technology and knowledge beyond our comprehension. According to believers, this hidden society may be responsible for various unexplained phenomena on the surface, ranging from UFO sightings to encounters with cryptids.

As we embark on our exploration of the Hollow Earth Theory, it is essential to approach these claims with both curiosity and skepticism. We will delve into the evidence and narratives put forth by proponents of this theory, carefully evaluating their validity and considering the implications of a world hidden beneath our feet. Our journey into the depths of this enigma begins with an open mind, a commitment to critical thinking, and a thirst for understanding the mysteries that may lie below.

The Hollow Earth Theory, with its beguiling notion of a subterranean world concealed within our planet, is no mere product of contemporary imagination; its roots run deep into the annals of human thought. To appreciate the evolution of this theory, we must journey back through the corridors of history to its earliest origins. Ancient civilizations such as the Greeks and Hindus entertained the idea of a hollow Earth, imagining vast inner realms inhabited by diverse beings.

However, it was in the 17th century that the Hollow Earth Theory began to take on a more structured form. The works of scientific luminaries like Edmond Halley, who suggested that the Earth might consist of concentric spheres, laid the groundwork for a more systematic exploration of this concept. Halley's proposal was grounded in the observation of variations in the Earth's magnetic field.

In the 19th century, the theory gained further prominence, and authors like Jules Verne captured the public's imagination with literary works such as "Journey to the Center of the Earth." These imaginative narratives contributed to a popular fascination with the concept of a hollow world, blending science with fiction in a manner that continues to captivate readers and thinkers to this day.

Throughout the 19th and 20th centuries, various proponents of the Hollow Earth Theory emerged, each offering their unique interpretations and claims. From Cyrus Teed's Hollow Earth Society to the writings of Richard Shaver, this theory found diverse expressions in literature and fringe beliefs. As we delve into the historical origins and evolution of the Hollow Earth Theory, we will uncover the fascinating tapestry of thought that has woven together elements of science, fiction, and speculation to create a theory that continues to intrigue and confound.

The allure of the Hollow Earth Theory extends far beyond the realm of scientific speculation and into the vibrant landscape of literature, art, and popular culture. It has inspired imaginative narratives that take us on captivating journeys to the subterranean realms beneath our feet. From the pages of Jules Verne's "Journey to the Center of the Earth" to contemporary works of science fiction, this theory has left an indelible mark on our storytelling traditions.

In Verne's classic novel, readers are transported to a world hidden beneath the Earth's crust, where intrepid explorers descend through labyrinthine caves, encountering extraordinary landscapes, creatures, and civilizations along the way. The novel captures the essence of adventure and discovery that permeates the Hollow Earth concept, inviting readers to suspend disbelief and embark on a voyage to the unknown.

The influence of the Hollow Earth Theory is not limited to literature alone. It has found expression in visual arts, with artists depicting subterranean landscapes, cavernous realms, and awe-inspiring vistas hidden beneath the Earth's surface. Paintings, illustrations, and digital art have all contributed to the rich tapestry of imagery associated with this theory.

Furthermore, the concept of a Hollow Earth has permeated popular culture, making appearances in films, television series, and video games. These adaptations often take creative liberties with the theory, using it as a springboard for fantastical adventures that captivate audiences and blur the lines between reality and fiction.

As we navigate the intriguing waters of the Hollow Earth Theory, we inevitably encounter the need for critical examination and scientific scrutiny. At the heart of this theory lie claims that challenge our understanding of the Earth's composition and structure, raising profound questions about its feasibility in the light of established geological knowledge.

The assertion that our planet possesses a hollow interior, hosting an advanced civilization and an inner sun, stands in stark contrast to the prevailing scientific consensus. Geology, the study of the Earth's structure and processes, provides us with a comprehensive framework for understanding our planet's composition. It tells us that beneath the Earth's surface lies a complex and layered structure comprising the crust, mantle, outer core, and inner core, each with distinct properties and characteristics.

Seismic studies, which involve the analysis of seismic waves generated by earthquakes, have offered us invaluable insights into the Earth's interior. These waves provide a window into the planet's deep layers, revealing a solid, dense core at the center. Gravity measurements, too, corroborate our understanding of a solid Earth, as they are consistent with the presence of a massive inner core.

Furthermore, the theory of plate tectonics, a cornerstone of modern geology, elucidates the dynamic processes that shape our planet's surface. It explains the movement of Earth's lithospheric plates and the formation of continents, mountains, and ocean basins. The Hollow Earth Theory, with its hollow interior, stands in stark contradiction to these established principles.

In our quest for understanding, we must recognize that the Hollow Earth Theory faces fundamental flaws and inconsistencies when assessed against the backdrop of geological realities. The weight of empirical evidence, derived from seismic studies, gravity measurements, and the principles of plate tectonics, points unequivocally toward a solid Earth with a layered structure. This scientific refutation prompts us to engage in critical thinking, discerning between speculation and the empirical foundations of geological science.

The Hollow Earth Theory, with its captivating narrative of a hidden world beneath our feet, beckons us not only through its audacious claims but also through the psychological allure it exerts on the human psyche. To understand why some individuals are drawn to embrace this unconventional belief, we must delve into the intricate realm of psychology and examine the factors that contribute to the theory's appeal.

One notable aspect of the Hollow Earth Theory's appeal lies in its capacity to offer a sense of wonder and adventure. It taps into the innate human curiosity about the unknown, kindling the flames of exploration within our minds. Just as Jules Verne's novels transported readers to uncharted territories, the Hollow Earth narrative invites us to embark on a mental journey into unexplored realms, igniting our imaginations in the process.

Furthermore, this theory often thrives on the allure of exclusivity and hidden knowledge. Believers may perceive themselves as part of an enlightened minority, privy to secrets that elude the broader populace. This sense of belonging to a select group can foster a sense of identity and purpose, reinforcing belief in the face of skepticism.

The psychological appeal of the Hollow Earth Theory extends beyond intellectual fascination. It can offer solace or escapism in a complex and sometimes uncertain world. Embracing such a belief can provide a sense of order and meaning, framing the world in a comprehensible narrative that soothes anxieties or resolves existential questions.

However, it is crucial to recognize that while the psychological appeal of the Hollow Earth Theory may be profound for some, it does not validate the theory's claims. It highlights the human capacity for imaginative exploration and the deep-seated desire to uncover hidden truths, even when these truths exist in the realm of speculation.

Conclusion

Our journey through the labyrinthine realms of the Hollow Earth Theory draws to a close, and as we emerge from its depths, we find ourselves standing at the crossroads of fact and fiction, science and speculation. This chapter has been a voyage through the corridors of an enduring enigma—a theory that challenges our understanding of the Earth's composition and kindles the fires of imagination.

In our exploration, we have encountered the audacious claims at the heart of the Hollow Earth Theory—the belief in a hollow Earth with an inner sun and advanced civilizations concealed within. We have traced the theory's historical origins, journeyed through the imaginative narratives it has inspired, and subjected its claims to the scrutiny of scientific refutation.

While the Hollow Earth Theory may continue to captivate the human imagination, we must ultimately recognize that it stands in stark contrast to the empirical evidence and established principles of geology. Seismic studies, gravity measurements, and the science of plate tectonics provide a compelling foundation for our understanding of the Earth's solid, layered structure.

Yet, as we conclude our exploration, we do so with an appreciation for the enduring appeal of mystery and the human desire to explore the unknown. The Hollow Earth Theory reminds us that even in the face of scientific refutation, the realms of speculation and wonder continue to beckon us.

As we navigate the complex terrain of extraordinary claims and unconventional beliefs, we carry with us the principles of critical thinking, skepticism, and scientific inquiry. These tools empower us to discern between fact and fiction, between the known and the unknown. In the grand tapestry of human thought and imagination, the Hollow Earth Theory serves as a testament to our unending quest for understanding, our boundless capacity for wonder, and our enduring fascination with the mysteries that lie just beyond the horizon.

Chapter 9: Genetically Modified Organism (GMO) Dangers - Separating Fact from Fiction

In the vast landscape of conspiracy theories, one subject that has sparked intense debate and generated a plethora of misinformation revolves around Genetically Modified Organisms (GMOs). While legitimate concerns and discussions surround the use of GMOs in agriculture and biotechnology, this chapter delves into the conspiratorial narratives that falsely claim these organisms pose dire threats to human health and the environment. Our journey through the world of GMO conspiracy theories will not only explore the nature of these claims but also provide historical context, scientific insights, and a reflection on the profound impact of misinformation. It is a journey that ultimately underscores the importance of informed discourse in navigating the complex terrain of GMO technology.

GMOs, organisms whose genetic material has been altered using genetic engineering techniques, have become a focal point of scientific advancement and public discourse. As we delve into this topic, we must acknowledge the legitimate debates surrounding GMOs, from questions about their economic and ecological impact to concerns about labeling and transparency. These debates are grounded in the complexities of modern agriculture and biotechnology, and they are crucial for informed decision-making.

However, beyond the realm of valid concerns lies a shadowy world of conspiracy theories that have leveraged the uncertainties surrounding GMOs to weave narratives of impending health disasters, environmental devastation, and corporate malevolence. These theories, often devoid of scientific rigor, rely on fear, misinformation, and pseudoscience to propagate alarming scenarios that mislead the public and influence policy decisions.

As we embark on our exploration of GMO conspiracy theories, we are guided by a commitment to separate fact from fiction, to engage in critical thinking, and to seek the truth in a realm where misinformation runs rampant. Through a comprehensive examination of the claims, historical context, scientific realities, the role of misinformation, and the importance of informed discourse, we endeavor to shed light on the multifaceted world of GMOs and the conspiracies that have cast shadows upon it.

At the heart of the controversy surrounding Genetically Modified Organisms (GMOs) are the claims that these organisms harbor hidden dangers, posing risks to both human health and the environment. This subchapter takes us into the heart of these conspiracy theories, exploring their specific claims and the narratives they propagate. From allegations of GMOs causing a wide range of health issues to dire predictions of ecological disaster, the spectrum of claims surrounding GMO dangers is vast and often alarmist.

Proponents of GMO conspiracy theories argue that these genetically engineered organisms, often found in common food products, are responsible for a host of health problems, including allergies, cancer, and even infertility. They contend that the genetic modifications introduced into these organisms, typically designed to enhance traits like pest resistance or crop yield, have unintended and harmful consequences for those who consume them. Additionally, some theories suggest that GMOs are responsible for environmental devastation, from the contamination of natural ecosystems to the decline of pollinators like bees.

Understanding the nature of these claims is essential as we navigate the complex landscape of GMO discourse. While it is crucial to address legitimate concerns regarding GMOs, it is equally vital to critically examine the sensationalized narratives that have emerged. This subchapter is our gateway into this intricate world of claims and counterclaims, where we will embark on a quest for clarity and understanding amid the haze of misinformation and fearmongering.

To comprehend the controversies surrounding Genetically Modified Organisms (GMOs) and the conspiracy theories that have taken root, it is essential to delve into the historical context in which GMOs emerged. This subchapter takes us on a journey through the annals of scientific innovation and agricultural evolution, highlighting the key milestones that led to the development and widespread adoption of GMO technology.

The story of GMOs begins in the latter half of the 20th century when scientific advancements in genetic engineering offered unprecedented opportunities to manipulate the genetic material of organisms. These breakthroughs paved the way for the creation of genetically modified crops designed to exhibit specific traits, such as resistance to pests or tolerance to herbicides. These traits promised to address longstanding agricultural challenges, from increasing crop yields to reducing the need for chemical pesticides.

As GMO technology gained traction, it stirred both optimism and apprehension. Proponents lauded its potential to revolutionize agriculture and alleviate global food shortages. Conversely, concerns emerged regarding the ecological and health implications of this technology. These concerns gave rise to legitimate debates about GMO safety and the need for regulatory oversight.

To comprehend the controversies surrounding Genetically Modified Organisms (GMOs) and the conspiracy theories that have taken root, it is essential to delve into the historical context in which GMOs emerged. This exploration takes us on a journey through the annals of scientific innovation and agricultural evolution, highlighting the key milestones that led to the development and widespread adoption of GMO technology.

The story of GMOs begins in the latter half of the 20th century when scientific advancements in genetic engineering offered unprecedented opportunities to manipulate the genetic material of organisms. These breakthroughs paved the way for the creation of genetically modified crops designed to exhibit specific traits, such as resistance to pests or tolerance to herbicides. These traits promised to address longstanding agricultural challenges, from increasing crop yields to reducing the need for chemical pesticides.

As GMO technology gained traction, it stirred both optimism and apprehension. Proponents lauded its potential to revolutionize agriculture and alleviate global food shortages. Conversely, concerns emerged regarding the ecological and health implications of this technology. These concerns gave rise to legitimate debates about GMO safety and the need for regulatory oversight.

Understanding this historical context is crucial as we navigate the complex landscape of GMO discourse, distinguishing between the promise of scientific advancement and the fears fueled by conspiracy theories.

Amid the swirling debates and conspiracy theories surrounding Genetically Modified Organisms (GMOs), a critical anchor emerges in the form of scientific consensus and empirical evidence. This subchapter delves into the heart of the matter—GMO safety—providing an in-depth analysis of the scientific realities that underpin our understanding of genetically modified organisms.

Over decades of research, GMOs have undergone rigorous testing and scrutiny to evaluate their safety for consumption and environmental impact. Scientific authorities and regulatory agencies around the world have established stringent protocols to assess the risks associated with GMOs. These protocols encompass comprehensive toxicity studies, allergenicity assessments, and evaluations of potential environmental consequences.

The overarching scientific consensus, supported by a multitude of peer-reviewed studies, is that GMOs currently on the market are safe for consumption and pose no greater risk to human health than conventionally bred crops. This consensus extends to the conclusion that GMOs have not shown to be inherently allergenic or harmful. Additionally, thorough assessments indicate that GMOs, when managed correctly, have no greater adverse environmental impact than conventional crops.

However, it is essential to recognize that the safety of GMOs hinges on proper regulatory oversight, adherence to best practices in agricultural management, and continued scientific vigilance. This subchapter will unveil the layers of scientific scrutiny and research that have contributed to the consensus on GMO safety, emphasizing the critical role of evidence-based decision-making in a world often clouded by sensationalized narratives and conspiracy theories.

In the digital age, where information flows freely and instantaneously, the impact of misinformation cannot be overstated. This subchapter delves into the role of misinformation, pseudoscience, and fear-based narratives in shaping public perception and discourse surrounding Genetically Modified Organisms (GMOs).

Conspiracy theories and false claims about GMOs have thrived in the fertile ground of the internet and social media. Fear-inducing narratives, often devoid of scientific rigor, have been amplified through viral content, leading to widespread confusion and apprehension among the public.

These narratives often tap into primal fears related to health, the environment, and corporate control of the food supply.

One consequence of misinformation is the erosion of public trust in scientific authorities and regulatory bodies. This mistrust can result in misguided policy decisions, public resistance to GMO adoption, and a distorted perception of risk. It is not uncommon for individuals to rely on unverified sources or anecdotal evidence, further fueling the spread of conspiracy theories.

Understanding the psychology of fear and the mechanisms through which misinformation spreads is vital in combating the impact of conspiracy theories. This subchapter will explore the ways in which misinformation can shape public perception, influence policy, and sow seeds of doubt in the face of scientific consensus. It underscores the importance of fostering media literacy, critical thinking, and evidence-based decision-making to navigate the turbulent waters of GMO discourse in the information age.

As we navigate the complex terrain of Genetically Modified Organisms (GMOs), one theme emerges as paramount—the necessity of informed discourse. This subchapter underscores the value of open and informed dialogue when addressing the multifaceted issues surrounding GMO technology.

Informed discourse begins with an acknowledgment of the complexities inherent in GMO technology. It recognizes that legitimate debates exist regarding GMOs, encompassing concerns about environmental impact, agricultural practices, and consumer rights. These debates are grounded in genuine questions about the ethical, ecological, and economic implications of genetic engineering.

Crucially, informed discourse hinges on evidence-based decision-making. It acknowledges the wealth of scientific research that supports the safety of GMOs currently on the market. It recognizes the importance of rigorous testing, regulatory oversight, and ongoing monitoring to ensure GMO safety. It distinguishes between the evidence-backed consensus on GMO safety and unfounded conspiracy theories.

Promoting informed discourse also necessitates an understanding of the ethical considerations surrounding GMOs. It acknowledges that the adoption and regulation of GMO technology involve complex trade-offs and requires transparency, accountability, and public participation. It emphasizes the need for responsible innovation and the ethical stewardship of biotechnology.

Ultimately, informed discourse is a call to action, urging us to engage in constructive conversations, grounded in facts and empathy, rather than fear and misinformation. It encourages us to embrace the complexities of GMO technology, recognizing that solutions to global challenges like food security and sustainability require thoughtful, informed, and evidence-based decision-making.

Conclusion

Our exploration of the realm of Genetically Modified Organisms (GMOs) and the conspiracy theories that have enveloped them draws to a close. This subchapter serves as a culmination of our journey—a journey that has taken us through the claims, historical context, scientific realities, the impact of misinformation, and the importance of informed discourse.

In concluding our exploration, we affirm that GMOs are neither a monolithic menace nor a panacea. They are a tool, a product of scientific innovation with the potential to address pressing global challenges in agriculture and beyond. However, like any tool, their impact depends on how they are used, regulated, and integrated into broader systems.

While legitimate debates persist regarding GMOs, evidence-based research and rigorous safety assessments have established that GMOs currently on the market are safe for consumption and have no greater environmental impact than conventional crops. The scientific consensus underscores the importance of separating fact from fiction, grounding our discussions in empirical evidence rather than fear-based narratives.

In navigating the complex landscape of GMOs, we recognize that responsible innovation and ethical stewardship are essential. GMO technology must be evaluated within the broader context of global challenges, including food security, sustainability, and ecological conservation. Informed discourse, underpinned by media literacy, critical thinking, and transparency, is the cornerstone of constructive decision-making.

As we step away from the intricate web of conspiracy theories and misinformation that often obscures the GMO discourse, we do so with a call to action—to engage in open, informed, and empathetic conversations. It is through such discourse that we can address the multifaceted issues surrounding GMOs, fostering a responsible and evidence-based approach to technology and innovation in an ever-evolving world.

Chapter 10: 5G Network Fears — Separating Fact from Fiction

Welcome, curious explorers, to a chapter that delves into the enigmatic world of 5G technology and the fears and concerns that have woven themselves into its narrative. In our ever-connected world, where the speed of information travels as swiftly as the signals that enable it, the advent of 5G networks has sparked both innovation and trepidation. In this chapter, we embark on a quest to unravel the conspiracy theories that allege 5G technology poses health risks, despite the absence of scientific evidence substantiating such claims.

5G technology represents a technological leap that promises faster data speeds, lower latency, and transformative possibilities for industries ranging from healthcare to transportation. However, amid this promise lies a shadowy realm of fears and concerns that have manifested in the form of conspiracy theories. These theories propagate alarming narratives, alleging that 5G technology emits harmful radiation, causes diseases, or poses risks to human health and the environment.

Our journey through the world of 5G network fears is a quest for clarity and understanding in the midst of sensationalized narratives and unverified claims. It is a journey that will lead us through the claims themselves, the historical context of 5G technology, the scientific realities underpinning its safety, the impact of misinformation, and the paramount importance of evidence-based discourse.

As we navigate this complex terrain, we do so with the conviction that technology, innovation, and progress must be balanced with a commitment to truth, facts, and responsible communication. The chapters that follow will serve as beacons of illumination, casting light upon the conspiracy theories that have entangled themselves with 5G technology, and guiding us toward an informed understanding of the potential benefits and concerns associated with this transformative innovation.

At the heart of the conspiracy theories surrounding 5G technology are claims that this innovation poses severe health risks to individuals and the environment. This subchapter delves into the specifics of these claims, offering a detailed exploration of the allegations that have fueled fears of radiation exposure, the spread of diseases, and other health-related concerns.

Conspiracy theories asserting 5G health risks often center on the notion that the electromagnetic radiation emitted by 5G networks is harmful. These claims argue that the increased frequency and proximity of 5G transmitters, required for the technology's high-speed data transmission, result in dangerous radiation levels. Some theories even go so far as to suggest that 5G technology is responsible for a range of health problems, including cancer, neurological disorders, and respiratory ailments. Additionally, unfounded narratives connect 5G networks to the spread of viral diseases, particularly drawing attention to the emergence of COVID-19.

As we delve into these claims, it becomes clear that they rely on a fundamental misunderstanding of the electromagnetic spectrum and the differences between ionizing and non-ionizing radiation. While 5G technology operates in the radiofrequency (RF) part of the spectrum, it emits non-ionizing radiation, which lacks the energy to ionize atoms or molecules and, therefore, cannot damage DNA or cells. Additionally, the scientific consensus has consistently found that 5G technology, when deployed within established safety limits, poses no substantiated health risks.

Understanding the nature of these allegations is critical as we navigate the debate over 5G technology and its potential health consequences. While it is important to address valid concerns and perform continuous research, it is also critical to distinguish between evidence-based evaluations and unjustified anxieties. This subchapter provides the basis for our investigation of 5G network concerns by giving light on the various allegations that have captured public attention and spurred conspiracy theories.

To comprehend the fears and concerns that have arisen regarding 5G technology, it is imperative to delve into the historical context in which this innovation emerged. This subchapter takes us on a journey through the annals of wireless communication, highlighting the key milestones that led to the development and widespread deployment of 5G networks.

The story of 5G technology begins with the evolution of wireless communication networks, from the early days of analog cellular systems to the advent of digital technologies like 3G and 4G. These predecessors laid the foundation for the development of 5G, with each generation building upon the capabilities and shortcomings of the last.

The need for 5G technology arose from the increasing demand for high-speed internet connectivity, particularly in an era where smartphones, smart devices, and the Internet of Things (IoT) have become integral to daily life. With the promise of faster data transmission, lower latency, and increased network capacity, 5G technology held the potential to revolutionize industries, from healthcare and transportation to entertainment and manufacturing.

However, it is critical to recognize that the advent of new technologies, particularly those involving wireless communication, frequently causes fear and suspicion. As 5G technology became more widely available, concerns about its possible influence on human health and the environment arose.

Knowing the historical context provides critical insights into the motivations driving the creation of 5G technology, as well as the first hopes and concerns around its adoption. As we progress through this chapter, it becomes evident that, while technological innovations have enormous promise, they also require critique and bring difficulties that necessitate scientific investigation and informed debate.

In the realm of 5G network fears, it is crucial to confront the scientific realities that underpin our understanding of the technology's safety. This subchapter delves into the extensive body of research, regulatory measures, and safety standards that have been established to evaluate the safety of 5G technology and its potential impact on human health and the environment.

Over decades, scientific authorities and regulatory bodies worldwide have established rigorous protocols to assess the safety of wireless communication technologies, including 5G networks. These protocols encompass comprehensive evaluations of electromagnetic radiation exposure, focusing on radiofrequency (RF) emissions generated by wireless devices and transmitters.

The overarching scientific consensus, supported by extensive peer-reviewed research, is that 5G technology, when deployed within established safety limits, poses no substantiated health risks to the general population. This consensus extends to the understanding that 5G networks emit non-ionizing radiation, which lacks the energy required to ionize atoms or molecules, and therefore cannot induce DNA damage or cellular harm.

Moreover, regulatory agencies such as the Federal Communications Commission (FCC) in the United States and equivalent organizations globally have established safety standards and exposure limits to protect the public from excessive RF radiation. These standards are founded on a wealth of scientific evidence and undergo regular review to ensure they remain up to date and protective of public health.

In the digital age, the dissemination of information and misinformation occurs at an unprecedented pace. This subchapter delves into the role of misinformation, pseudoscience, and fear-based narratives in shaping public perception and discourse surrounding 5G technology and its alleged health risks.

In the linked world of social media and the internet, conspiracy theories and misleading claims concerning 5G technology have found fertile ground. Fear-inducing tales, typically lacking of scientific rigor, have been circulated via viral content, causing considerable public misunderstanding and worry. These stories play on primordial worries about health, the environment, and the unknown.

One of the consequences of misinformation is the erosion of public trust in scientific authorities and regulatory bodies. This mistrust can result in misguided policy decisions, public resistance to 5G technology deployment, and a distorted perception of risk. Individuals may turn to unverified sources, anecdotal evidence, and echo chambers, further amplifying conspiracy theories and misinformation.

Among the turbulent seas of 5G network worries, one guiding idea stands out: the necessity of evidence-based debate. When addressing the various difficulties surrounding 5G technology, health concerns, and the rise of conspiracy tales, this subchapter emphasizes the importance of open and educated discourse.

Informed debate begins with an acknowledgement of the intricacies inherent in 5G technology, as well as the reasonable concerns raised about its adoption. It acknowledges that responsible innovation needs continual scientific investigation and evaluation, as well as open communication with the public.

Crucially, evidence-based discourse hinges on the recognition of the scientific consensus that 5G technology, when adhering to established safety limits, poses no substantiated health risks. It acknowledges the role of regulatory measures and safety standards in ensuring the safety of wireless communication technologies. It distinguishes between evidence-backed assessments and unfounded fears.

Promoting evidence-based discourse also entails fostering media literacy and critical thinking. It empowers individuals to navigate a world awash with information, misinformation, and conspiracy narratives. It encourages open and respectful conversations that consider the complexities of 5G technology, its potential benefits, and the legitimate concerns that require attention.

Finally, evidence-based discourse is a call to action, encouraging us to have productive talks based on facts and empathy rather than fear and disinformation. It recognizes the need of balancing technology and innovation with a dedication to truth, openness, and responsible communication. As we negotiate the complex web of conspiracy theories and 5G network worries, evidence-based discourse serves as a guidepost, directing us toward a more informed and equitable future.

Conclusion

Our journey through the enigmatic realm of 5G network fears draws to a close, leaving us with profound insights into the intersection of technology, fear, and misinformation. This subchapter serves as a culmination of our exploration, affirming the significance of scientific consensus, regulatory measures, and evidence-based discourse in assessing the concerns and potential benefits associated with 5G technology.

As we step away from the tumultuous waters of conspiracy theories and fears, we do so with a call to action—a call to engage in open, informed, and empathetic conversations about the transformative power of technology. It is through such discourse that we can address the multifaceted issues surrounding 5G technology, fostering responsible innovation and a society that navigates the complexities of progress with clarity, reason, and an unwavering commitment to truth.

Closing Word

Dear Explorers of Truth,

As we conclude this captivating journey through "Unraveling Mysteries: Exploring Science and Debunking Myths - Storytelling for the Whole Family," encompassing both Volume One and Volume Two, I trust that it has been as exhilarating for you as it has been for me. Together, we've delved into the intriguing territories of scientific inquiry, debunking myths surrounding topics such as a flat Earth, vaccine misinformation, alien cover-ups, HAARP weather control theories, GMO controversies, and 5G technology debates. Each myth, a puzzle in its own right, has provided us with a platform to engage with science in a manner that is both captivating and enlightening.

This comprehensive work was crafted not merely to dispel misconceptions but also to foster a spirit of inquiry and critical thinking. By dissecting these myths, we have gained a deeper comprehension of the world around us and the significance of evidence-based reasoning. It's about peering beneath the surface, questioning the unknown, and seeking truth in the vast sea of information.

I hope these narratives have sparked your curiosity and armed you with the knowledge to grasp and elucidate these intricate subjects. Keep in mind that the pursuit of knowledge is an unending odyssey, and each stride we take in comprehending the world around us propels us toward a brighter, more informed future.

As we bring this chapter to a close, I implore you to carry forward the spirit of inquiry and the delight of discovery. Let the lessons and tales from this compendium serve as a guiding light in your lifelong expedition of learning and exploration.

I extend my heartfelt gratitude for accompanying me on this riveting journey. May your path forever be illuminated by the radiance of knowledge and the joy of discovery.

www.ingramcontent.com/pod-product-compliance
Lightning Source LLC
Chambersburg PA
CBHW062236290526
45794CB00006B/2310